AMERICAN LEVEL PATENTS
ILLUSTRATED AND EXPLAINED

VOL. I, NEW ENGLAND

DON ROSEBROOK

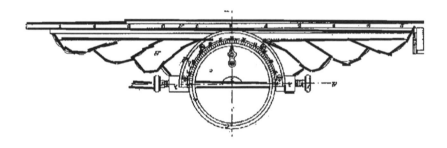

ASTRAGAL PRESS
MENDHAM, NEW JERSEY

International Standard Book Number: 1-879335-92-1
Library of Congress Catalog Card Number: 00-101308

Published by
THE ASTRAGAL PRESS
5 Cold Hill Road, Suite 12
P.O. Box 239
Mendham, NJ 07945-0239

Manufactured in the United States of America

DEDICATED TO ALL LEVEL COLLECTORS

Those who have found something different and

wondered what its origin was.

Look here for information about the tool

and the man who invented it.

ACKNOWLEDGMENTS

I wish to express my great appreciation for Lars Larsen and his continuing effort to locate patents pertaining to levels. He labored long and faithfully to identify and copy the patents discussed in this book, and many other patents pertaining to levels or to level accessories. This book, and others yet to come, will distribute to all some of the fruits of Lars' work.

Another grateful thank-you is due my wife, Pam, for her assistance in locating information on the patentees and for her companionship on the many long driving trips to libraries around New England. Her encouragement has been invaluable and her tolerance of my addiction to this task has been wonderful.

Thanks to Ken Thomas for his review of the introduction to this volume and for comments pertaining to the process of obtaining a patent.

Thanks to Mel Miller for discussions and copies of some patents and to Jim Mau for copies of the photographs for the Davis Design Patents of May 1877.

Thanks are also due to Martyl and Lisa Pollak, at Astragal Press, for their caring assistance in getting this material into print.

FOREWORD

There are many patents of interest to collectors of levels and of accessories useful to users of levels. This book, a compilation of New England patents, is a companion to *American Levels and Their Makers, Volume I, New England.* Other companion books will follow as subsequent volumes of *American Levels* are published.

In the many library visits necessary to collect information on the makers of levels, it seemed only proper to collect information on the patentees of those levels and of levels possibly never manufactured. After all, some of those levels might turn up some day, and the maker and the patentee might be one and the same person. (That has already proved to be true.) At the conclusion of the library research (at least as far as I was going to go anyway), it seemed a shame to hoard the information that I had gained. Just publishing my notes, however, would not suffice.

What else might be appropriate to accompany the notes was a puzzle. It seemed impractical to publish the entire text of all of the pertinent patents, yet to reduce the information to the presentation of a patent drawing by itself seemed too little. If there was to be a book, it must be worthwhile. Patent drawings were sometimes very misleading and I often couldn't deduce the actual intent of the invention without reading the patent (sometimes reading the patent many times). Prior to this research, in fact, I had often wondered if patented articles were faithful to the concept proposed in the patent description.

So there coalesced the pertinent parts of a book. Who dunnit? What did they think they were doing? What did they actually do? That's the way this book is designed. A page (sometimes a few more) is devoted to each patent. On that page you will find, in the heading, the patentee and the date, his city of residence, the patent number, and the title of the patent. Each page starts with a presentation of what is known about the patentee, i.e., who dunnit?

This is followed by a brief abstract of the patent, i.e., what did they think they were doing? The last items on a page consider whether or not there were any known tools produced according to the concepts of the patents, and whether or not they incorporated all of the concepts of the patent, i.e., what did they actually do? Finally, each page contains a copy of the patent drawing, large enough to contain all of the elements of the patent in a size that allows interpretation.

I must stress that I am not an expert on materials of construction, mechanical engineering, or on terms used by machinists to refer to parts of machines. Journal, for instance, was a new word for me. I found it to be defined as that part of a rotary shaft that turns in a bearing. There were other words and concepts that were not very clear to me. As a consequence, some of the words describing a patent may not be quite proper, but it is my hope that I have conveyed the intent of the patent so that, in combination with the patent drawing, the reader can understand the patent.

Likewise, I am not a genealogist or trained historian. Thus, I was freed from the need to trace down every last detail about the person obtaining the patent and his environment. I have researched these men in order to obtain clues about whether or not a given device might ever have been produced. As will become apparent in subsequent books, I did discover the identity of manufacturers or individual makers of levels that I had seen but had not heretofore been able to assign to a specific manufacturer, and I did discover new makers by doing this research. I have decided to share the historical information as I found it.

Sometimes the patentees were mystery men. Here today and gone tomorrow. Especially in the mid-1800s, men went to search for gold, followed work on the railroads, fought in the Civil War, or died too young. Sometimes the patentees were men of great substance, i.e., bankers, doctors, and lawyers. These professions do not preclude them from being inventive, but my suspicion is that when they appear as co-patentee they contributed some financial support or other thing for which this patent was payment. Some patents were granted to persons who were listed as laborers or mill hands or other menial job holders. For many, this may have been an entry level job and should not be an indictment of their skills or imagination.

When it came to recognizing whether or not a manufactured level incorporated some or all of the concepts of a patent, I took apart many levels and compared them to the patent's dictates. If I found them to be significantly different, I tried to convey that information in this book.

It is my hope that if any readers desire to have more information about these men, they will find my notes to be a valuable starting point. If any collector finds a level that conforms to a patent that I have not yet seen, it is my hope that they will let me know and that we can eventually spread the word among all collectors.

This book has a relatively untried format and I hope that readers will enjoy and learn from its information.

Organization

This book is organized chronologically by patent date. At the end of the book, however, there are several indices organizing the material:
- by name of patentee
- by type of patent
- by patent number

The reader with just one or two patent clues, therefore, will be able to find the level he or she might be looking at.

INTRODUCTION

Sometimes, when we see a patent, we wonder why that "thing" deserved to be patented. Carrying that thought a little further leads to questions about how and why anything gets patented. So perhaps it is appropriate here to look at the process for obtaining a patent, and its history. While we're looking there, perhaps some interesting questions will arise. Answers to those very interesting questions are not necessarily forthcoming, e.g., was the timing of the Patent Office fire, coincidental with the institution of much stricter regulations for granting a patent, just a convenient accident?

In the beginning, the individual colonies had the right to grant a patent. The first of these was granted in 1641. These patents generally followed English patent law. As a matter of fact, the principles of English patent law governed the granting of patents in the Colonies for another 100 years or more. Upon gaining independence from England, and establishing a government, the United States showed a continuing concern for innovators.

The Federal Constitution empowered Congress to make laws regulating patents: "...to promote the progress of science and useful arts by securing for limited times to ... inventors the exclusive right to their...discoveries." Looking at this we see key words like "limited times" and "exclusive right" and "discovery." The Constitution, thus, provided for a limited time a monopoly as a reward for an inventor's labors. The use of the word "discovery" meant that in order for the inventor to have a monopoly he must propose something new.

Congress first passed a patent law in 1790 and envisioned 14 years, with a seven year extension possible, as the appropriate time for exclusivity. The granting of a "letters patent" was to be discharged by the Secretary of State, the Secretary of War, and the Attorney General or any two of them. Thomas Jefferson was the Secretary of State at the time and he personally attended to the granting of the patents. This proved to be unwieldy and Congress rewrote the law in 1793. Under this legislation, the responsibility stayed in the State Department and a grant required the approval of the Attorney General, but little oversight was required or provided.

From 1793 to 1836, it was possible to obtain a patent merely by complying with the formal requirements of submitting a description of the invention plus a drawing, a model, and the appropriate fees. It takes little imagination to see what ensued under that system. There were many conflicts and, in many cases, patents were obtained for objects that were in common use or that had no novelty.

The patent law of 1836 addressed these problems by allowing the creation of a Patent Office. The office was to consist of a Commissioner, an Assistant Commissioner, and three Examiners-in-Chief appointed by the President, and several lesser employees nominated by the Commissioner and appointed by the Secretary. The office was made a sub-department of the State Department. Essentially, the rules were the same, but now they were enforced. Somehow, it seems appropriate that there was a fire at the Patent Bureau in 1836; it destroyed all the existing patent records, thus allowing folks to start over on equal footing under the new law. In 1847, when the Department of the Interior was formed, the Patent Office was put under its oversight (and subsequently under that of the Department of Commerce in 1925).

Patents could be denied on the grounds of prior use or lack of novelty, and quite probably for non-functionality. Laws established routes of appeal for the would-be patentees whose patents were denied. The Patent Office published regulations and guidelines for inventors. In 1861, the term of a patent was changed from the original 14 plus 7 years to 17 years. In 1870, the requirement for a model was removed and many other changes were made to patent law.

Some early inventors were damaged by ignorance of the difference between United States Patent Law and English Patent Law. The fatal difference was that the United States' patents were granted for novelty based on the invention being unknown or unused at the time of the *discovery* of the invention. English patents require that the invention be unknown or unused at the time of the *grant of the patent*. Thus, a device could be manufactured here and stamped with "patent applied for," and if the patent description was sufficient and the device was deemed to be novel and useful, then the patent was granted. But, by doing that, the inventor was ineligible for an English patent because the device had now been used (and manufactured) before the patent was granted. This meant that anyone could produce such a device in England without fear of prosecution for patent infringement.

Design patents could also be granted and several are contained in this volume. The term of a design patent could be 3½, 7, or 14 years at the option of the inventor and according to the size of the fee one was willing to pay. Patent applications had to be submitted in the same fashion as for a device or a process.

Some patents were indicated to be reissues. This was done for defective patents. Patents that have been granted may be found to be defective, because they either do not make a sufficient claim (to suit the patentee) or they make claims found to be excessive. Excessive claims are claims for things found to be in conflict with prior patents or found to be in common use. Reissue patents had to go through the same patenting process all over again, with a patent application and examination by the Patent Office Examiners. Some level patents of interest to us are reissue patents.

If a person, who obtained a patent, wanted to utilize the full protection granted by patent laws, it was necessary to mark the patented object with "patented" and the date of the patent. If the device was too small to contain that notation, then the package was required to be marked. Patented, but unmarked, devices were protected from *unknowing* infringement to the extent that the competing device had to be removed from the market, but the patentee could not obtain damages.

Marking an article as patented when it is not is prohibited by law. Marking an article either "patent pending" or "patent applied for" or use of similar wording was permitted and was considered to be only an information notice and had no force of law. Continuing to mark articles in that fashion after a patent was no longer pending, or if no application had been made, was prohibited by law.

Unless determined by some prearranged agreement, joint patentees had equal rights to the invention. That meant that each could independently manufacture and market the device, and each had the right to sell his share of the patent to whomever he pleased. Patent rights might be assigned to one of a group of co-patentees, or to another individual or group of individuals. In many industries it was normal for the company to demand that the employee assign his patent rights to the company. It was normal that the assignee had exclusive rights unless there were more than one of them and unless the grantee assigned only a portion of the rights.

The steps in obtaining a patent included submitting an application with a written description of the device or process. These documents had to be signed by the inventor and two witnesses. In addition, the submission had to include a drawing and, prior to 1870, a model. Apparently, for some time after 1870, a model could be required but it was not a general requirement. Models were submitted with some applications for many years after 1870, and many patents have a line beneath the title indicating the date that the application was submitted and whether or not a model was also submitted. This practice continued into the early 1900s.

The patent examiners then reviewed the submissions. The discovery had to be novel and useful. Novel meant that the discovery was new and not just unpatented. Commercial success was considered to be evidence of utility. Thus, sales of an item with the "patent applied for" status could be useful to an inventor for more than immediate monetary reasons. On the other hand, patent applications could be rejected for lack of novelty. Patent applications could also be rejected for interfering with another pending or expired patent. In this case, it was the responsibility of the patent examiner to determine which patent took precedence. Both patents were adjusted to contain the identical claim and both were considered to be novel. The examiner's task then was to determine which person had the earliest discovery. That person's claim was allowed while the other's claim was disallowed.

In the beginning, neither the inventor nor the witnesses signed the patent drawings. In the mid to late 1850s a transition occurred and many patent drawings were signed. By 1860, it appears that all patent drawing bore the inventor's signature along with that of the witnesses. By the late 1860s it became usual for the attorney to sign the drawings for the inventor. The attorney usually arranged to have the final drawings produced whether he signed them or not.

The patents described and discussed in this book, and subsequent books, fall into many categories. Some are design patents, some are revised patents, some are renewed patents, but most are just patents describing novel devices. Examination of levels will reveal many in-stances of them being marked with improper patent dates and/or vague dates that cannot be verified. Some levels are known to carry invalid patent dates, i.e., patents were always issued on Tuesdays and some levels are marked with dates that are not a Tuesday. There are other instances, such as Stanley's use of the date February 23, '92 on its Nº 98 and other levels. No applicable patent on that date is known to the author. Davis utilized a patent date on his wooden carpenters' levels that was exactly one year too early. Stanley used the date Sept. 69, but no patent can be found, and this may have been an incorrect stamp for an intended Sept. 67.

The reader will note with suspicion that some of the claims in patents do not appear to be novel, and that is indeed the case. It would appear to be a case of inattention by the examiner or perhaps ignorance of instances of prior usage. In any case arising from this, such as Traut's claim for a hexagonal level after it had been in use for over 10 years, that part of the patent would not be enforceable and would have been an embarrassment to Stanley.

A review of the level patents leads to the conclusion that a greater percentage of the devices were actually produced during the mid-1800s than later. Review of the patents reveals some other interesting facts. During the 1800s, patent descriptions were likely to consist of a single page (even though the device might be complex) and a patent drawing would consist of a single sheet. As the 20th century got underway, the language of patents seems to the layman to have become more complex and repetitive; drawings seem more numerous and more detailed. A special format arose for making patent claims, and even a simple concept such as Traut's patent of April 26, 1903, for a chambered vial gave rise to a large number of claims (22 in Traut's case and the general idea was not at all new). The reason for this was the continually increasing amount of patent law brought by Congressional and Judicial action. Today, the language of the claim must stand alone and not refer back to language in the descriptive part of the patent. Thus, independent claims are made for every potential application and the patent becomes very long and appears to be highly repetitive.

William G. Ladd, Jr.
Cambridge, Massachusetts

FLUID-LEVEL

William Garner Ladd, Jr. was born in 1824, and married Lucinda Ireland of Somerville in 1846. The Cambridge City Directories from 1848 to 1853 show them living on Main Street, north of Dana Hill. No occupation is shown. William G. Ladd, Sr. was also living at the same address and no occupation was provided for him either. William Ladd, Jr., however, was a partner in a hardware store in Boston. His partners were Charles W. and Samuel J. M. Homer. By 1861, he was involved in Ladd, Webster & Co., manufacturing sewing machines.

The patent is relatively broad, calling for a shallow cylindrical chamber, or a "tube in the shape of an entire ring," half filled with quicksilver (mercury) or other fluid, in combination with a graduated annular dial. A floating needle is optional.

This level was manufactured in a form similar to that dictated by the patent. It is marked only with "Wm. G. Ladd, Jr." The known example is mercury filled, and contains a scale, marked on ivory, at the back of the chamber. A small iron bar floats on the mercury.

The only changes from the patent design are the substitution of the ivory scale for the graduated annular ring and the absence of the ears on the retaining ring.

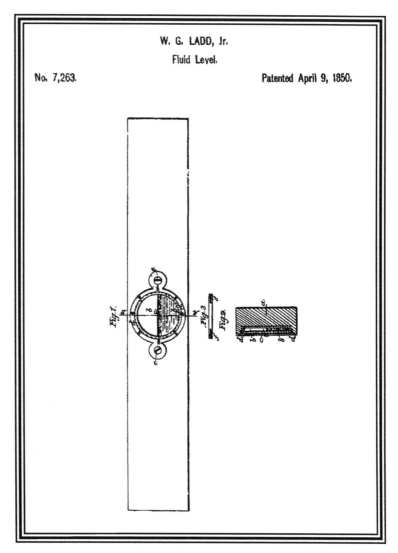

IMPROVED REFLECTING SPIRIT-LEVEL AND SQUARE

Francis Wilbur (sic) is listed in the Roxbury City Directories of 1850 and 1852, with carpentry as his occupation. The Vital Records of Roxbury show Francis Wilbar as a carpenter. No other information was found about him.

The patent describes a device consisting of a cube of wood, wherein two adjacent interior vertical sides are fitted with mirrors. The top is to contain two spirit vials placed at right angles to one another, with the long axis of each parallel to one of the mirrors. Wilbar claims that one of mirrored sides may be hinged to enable the user to lay off any desired acute or obtuse angle. He also claims that a spherical, or crowned disk, may be substituted for the two spirit vials.

The patent document contains a detailed example of the use of this device, both as a square and as a level, by a mason building a wall. Its accuracy and utility seem limited.

To date, no examples of Wilbar's patent have been identified.

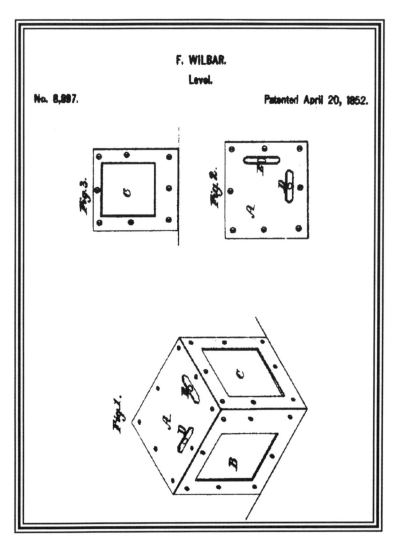

F. WILBAR.

Level.

No. 8,897.

Patented April 20, 1852.

SPIRIT-LEVEL

Lebbius Brooks lived in, what is now Somersworth, New Hampshire.[1] Information from the inventory indicates that Brooks was actively engaged in the manufacture of the levels at Somersworth, probably as early as 1853. He also has other patents for tools, including a saw set. He was born about 1830 and he died just prior to September 1, 1858. The patent rights were sold at auction after his death.

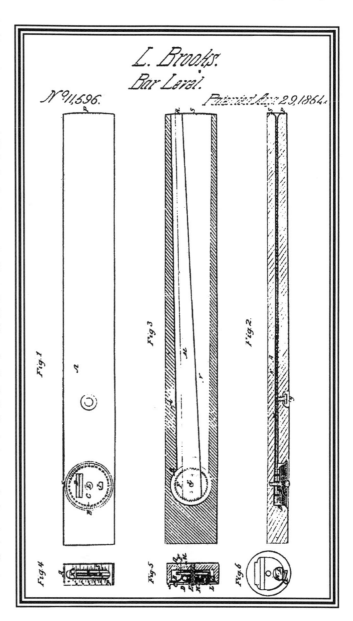

The device described in Brooks' patent did not include a horizontal level vial. He envisioned only one spirit vial and that was to be on the circular plate associated with the bar or pointer. The circular plate was to be divided into a number of divisions, and a tooth provided for each division. Five degree segments were suggested. The circular plate would be fixed by means of a spring catch. The spring catch was to be manually operated. The object was to release the catch and rotate the circular plate containing the level vial until the division nearest to a level indication was reached. Then, with the catch in place, the bar would be moved against a five degree scale on the end of the level. When the vial indicated level, the sum of the degrees on the wheel and the degrees on the end equaled the inclination. The level, as manufactured, does operate in the proposed manner.

Brooks called this the "bar level." Half of the patent rights to Brooks' level were owned by L. M. Tibbets. The Stanley Rule & Level Co. obtained the rights and sold this as the Brooks' Patent Universal Level (Nº. 26).

---

[1]   According to R. Smith (Stanley Collector News Vol.1 No.1 pp10-13) it was local custom to call the town Great Falls because of its location near the largest waterfalls on the Salmon Falls River.

CARPENTER'S RULE

Lorenzo Case Stephens (1809-1871) was a rule maker who, at one time, had worked for Hermon Chapin. In 1854, Lorenzo and his son, Delos Hart Stephens began their own rule making business in quarters leased from Chapin. In the beginning, Stephens acted as an independent contractor making rules for the Chapin firm.[2]

The single claim of the patent was for a rule with a moveable blade and spirit level attached.

This patent was used in the famous combined rule, square, level and bevel. The tool was manufactured by Stephens as the № 36. Subsequent to the merger with Chapin, the rule was manufactured by Chapin-Stephens Co. as № 036 and, subsequent to their acquisition by the Stanley Rule & Level Co., the № 036 was continued. Stephens also produced this rule in ivory with German silver fittings carrying № 38. The rule is also known in an ebony format. Stanley produced the tool only in boxwood with brass fittings. Metric scales on this rule were available by special order. The total manufacturing lifetime of the rule was 84 years. (It was discontinued in 1942.)

The tool is made of a double thickness of brass bound wood. All known examples are 12" two-fold rules with a square joint and a steel scale pinned at one end and free to pivot. The scale is marked to indicate angles and thus can be used as a protractor. The tool can also be used as a square and a bevel.

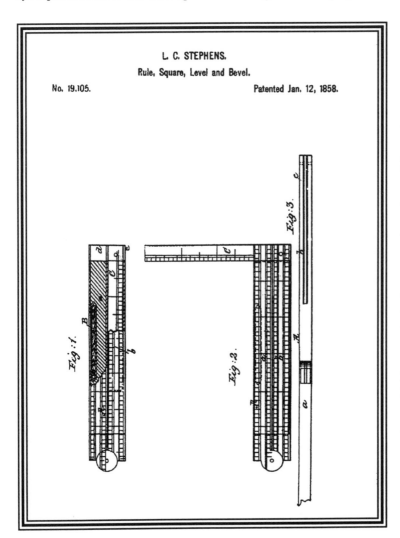

L. C. STEPHENS.
Rule, Square, Level and Bevel.

No. 19,105.

Patented Jan. 12, 1858.

[2] Kenneth D. Roberts, *Wooden Planes in 19th Century America, Volume II, Planemaking by the Chapins at Union Factory, 1826-1929.* Fitzwilliam, NH 1983. pp 162-165.

William T. Nicholson
Providence, Rhode Island

May 1, 1860
28,104

SPIRIT-LEVEL

William Nicholson is better known for making files than for levels. He was born in Pawtucket, Rhode Island on March 22, 1834. He worked for several years in machine shops and, in 1858, went into business for himself making jewelers' tools and light machinery (and levels as it turns out). He made levels, other than the type he patented, during this time; advertisements in 1860 show many of the shapes of levels that he made. During the Civil War he made parts of small arms, machine tools, and screw machines. In 1864, he founded the Nicholson File Co. He died in Providence, in October 1893, at age 59.

Nicholson's Patent is for the rolling, or rotatable, vial cover and the lower slots therein that enabled the user to read the level from below.

This level was made by Nicholson and then by Stanley, but Stanley used only the frame and did not include the revolving vial cover.

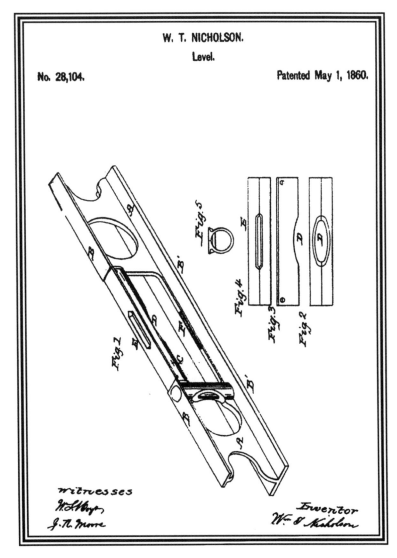

-5-

IMPROVEMENT IN COMBINED SPIRIT-LEVELS

The Cahoon family has been in Harwich since the late 1600s.[3] Alvin Jr. was born on March 25, 1812, one of nine children of James Jr. and Lettice Cahoon.[4] (Makes one wonder where the Alvin Jr name originated.) He was married in 1833 to Clarissa Young.

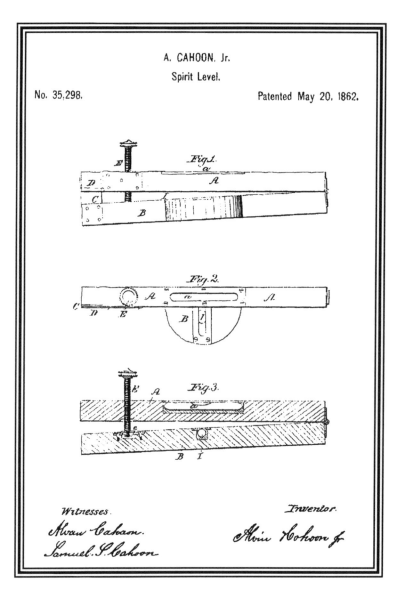

A. CAHOON, Jr.

Spirit Level.

No. 35,298.   Patented May 20, 1862.

Fig.1.

Fig.2.

Fig.3.

Witnesses.
Alvan Cahoon.
Samuel S. Cahoon

Inventor.
Alvin Cahoon Jr

Cahoon's patent claim is for "two horizontal spirit-levels at right angles one to the other, and one of them rendered adjustable in the vertical plane for the purpose of determining the trim of a vessel..." This patent is one of the few level patents that refers to a prior patent (or to an application) and, in this case, it is to the rejected application of Aston in 1856. In the drawing, *E* is a screw used to adjust the angle of the top level, and *C* is a graduated scale attached to the bottom level. *D* is a pointer or vernier scale and is attached to the top level. As shown in the drawing, the bottom level is widened in the center to accommodate the spirit vial.

No specific examples of a tool made according to this patent are known to the author. The device was meant to be used in determining the trim of vessels during loading. Cahoon calls it a "ploiometer" and that term is not in a standard dictionary. It appears to be similar to a grade level in some respects.

---

[3]   Paine, Josiah, *History of Harwich 1620 to 1800* (Reprint of 1971 by Parnassus Imprints of Yarmouth MA)

[4]   Vital Records of Harwich 1694-1850

IMPROVEMENT IN PLANE ANGULOMETERS

Eli Thayer was an educator and politician. He, along with other emigres from free states, was instrumental in the settlement of Kansas, thus keeping her from becoming a slave state. Later, in 1856, he was elected to Congress where he stayed until 1861. He was said to be responsible for the admission of Oregon to statehood. Under Lincoln, in 1861-1862, he was a special confidential agent of the Treasury Department. Throughout the 1860s, he was a resident of Worcester. He was considered to be an expert on matters of invention. He had an Master's degree from Brown University and had qualified for the bar, but never sought admission. He had been the Principal of Manual Labor High School (later Worcester Academy) and had founded the Oread Academy (a university type academy for women, the first of its kind). Thayer was a direct descendent of John and Priscilla Alden and was born in Mendon MA in 1819; he died in 1899.

The device pictured here is basically a weighted needle inclinometer that, by use of two perpendicular sets of knife edges, is free to rotate in each of two directions, thereby simultaneously giving the angles of a horizontal plane from the horizon.

Thayer's claim is for a "pendulum moving upon three or more bearings in the same plane and carrying upon its top a graduated arc, and its combination with the spherical surface and the opening therein..."

To date, no examples of this device have been identified. It would probably be considered as primarily a scientific instrument.

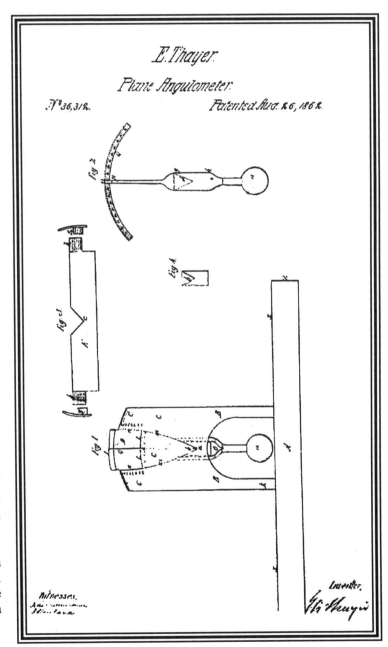

IMPROVEMENT IN COMBINED PLUMB AND LEVELS

Frederick Traut was the father of Justus Traut. Both he and Justus had worked for Hall & Knapp and both came to Stanley Rule & Level in the merger between A. Stanley and Hall & Knapp. This was his only patent for a level or leveling device.

The patent described the device as a right angle intersected at each end by a circular plate. A bar containing the spirit level was to be pivoted at the intersection of the two legs of the right angle and, thus, to have the free end be capable of moving along the inside edge of the circular plate. The free end of the bar was to be provided with a thumb screw to secure it in any desired position. The circular plate was to be graduated in degrees. A depression was to be cut into the stock to receive the device.

The patent for was used for the Stanley № 32 level. The inclinometer device could be put into any level, but Stanley sold the № 32 as its inclinometer level. The basic level used to make the inclinometer was the № 9. Over the years, the style of the № 32 changed as the style of the № 9 changed.

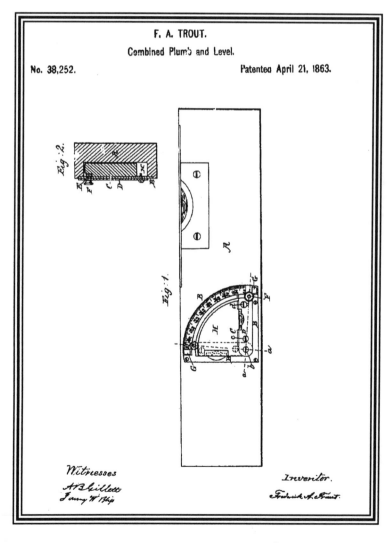

F. A. TROUT.

Combined Plumb and Level.

No. 38,252.   Patented April 21, 1863.

Fig. 2.

Fig. 1.

Witnesses
A B Gillett
Jenny W Phip

Inventor.
Frederick A. Traut.

The directions for the operation of Frederick Traut's device were contained on a green card inserted into the mortise behind the works of the inclinometer. The user had only to loosen the knurled thumb screw and pivot the arm carrying the level vial to the desired angle.

IMPROVEMENT IN SPIRIT LEVELS

H. S. Shepardson was a machinist and tool maker by trade. He is perhaps better known for his patented bit braces and various patented catches and latches. The H.S. Shepardson & Co of Shelburne Falls brought together several inventive tool makers in addition to Mr. Shepardson.

This tool contains a standard level spirit vial (in a vial coffin) in the metal frame. There is a second spirit vial attached to a graduated disk in the side that can be rotated so as to set the vial at any desired angle. The rotation is controlled with a thumb screw on the opposite side of the level from the spirit vial in the circular disk. The metal frame is fitted with thin wooden sides.

The tool also contains an adjustable internal sighting tube. This is best observed in *Figs. 2 & 4*. The adjustable sights are shown on the left hand side of *Fig. 3*.

The only claim for this level is that for a metal frame with thin wooden sides.

This seems like a practical device, but no example has yet been identified.

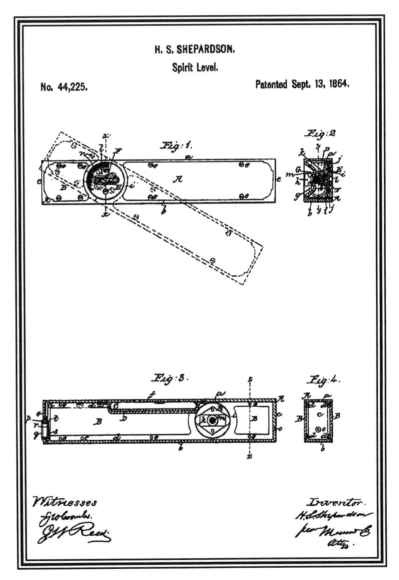

-9-

Aaron Chase Jr.
Somerville, Massachusetts

August 29, 1865
49,675

Assigned to himself and Timothy Howe of Somerville, MA

IMPROVEMENT IN INCLINOMETERS

Aaron Chase was listed in the 1869-70 Somerville, Arlington, and Belmont City Directory as a carpenter and a principal of Chase and Austin. In the 1870-71 Directory, he is listed only as a carpenter. But there is a full page ad in the directory for A and AB Chase, "Carpenters and Builders."

No information about Timothy Howe could be found.

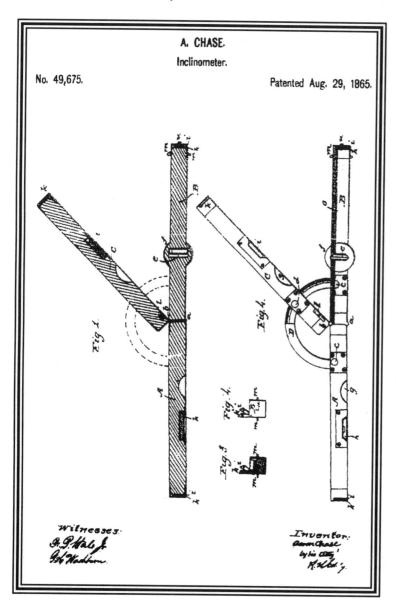

A. CHASE.
Inclinometer.
No. 49,675.
Patented Aug. 29, 1865.

Witnesses:

Inventor:
aaron Chase

The tool consists of three hinged bars (rule joints) such that they may be folded together into a flat piece. Each piece contains a spirit vial. The arc piece is detachable and goes through the top bar, and fixes into the other two, when they are extended end to end. Bar *B* carries a spirit vial in the "plumb position" while the other two carry level vials. Bars *A* and *B* carry sights on their extreme ends which may be rotated into position for use.

Bar *B* is fitted with spring catches to lock it to either or both of the other two bars.

At this date, no example of this tool has been observed, although English tools of somewhat similar design are known.

William L. Richardson
Reading, Massachusetts

January 1, 1867
60,788

Assigned to himself and F. E. Nutting of Florence, MA

IMPROVEMENT IN LEVELS

There is no record of Richardson in Reading, Mass. In South Reading, the absentee tax roles from 1850's show William L. Richardson of Boston paying taxes. In 1860, the heirs of William L. Richardson are on the tax roles for land and a shop. (The only William L. Richardsons in Boston in the 1850s and 1860s were physicians, with the senior Richardson dying in 1852.)

F. E. Nutting was, most probably, Freeman Nutting, a son of John Nutting of South Amherst. His brother, Porter, was the leading mason contractor of the North Hampton area. In 1891, Freeman was in Florence but he died before 1895. No information, contemporary with the patent, is available at this time.

The claims are for a device that has a plummet attached to the *center* of the ball joint near the base of the apparatus; that has detents on the vertical shaft to facilitate the attainment of an exact right angle; and, that has a detachable level, the top rail of which is a sighting tube, and the spirit vial of which can be turned 90° to act as either a plumb or a level.

There is a shaft extending upward from the nut over the ball joint, and fittings sitting upon that shaft may be rotated upon loosening set screw *a*. Once locked into place, the upper table may be rotated by loosening the set screw *b* that fits into grooves in the shaft of the table.

The detachable level contains a sighting tube along its upper edge and a spirit vial attached to a circular disk that is held in position by a spring. Detents on the disk allow it to be rotated exactly 90°, so that it may be used as a plumb.

To date, no examples of this device, either with or without the mounting part, have been observed.

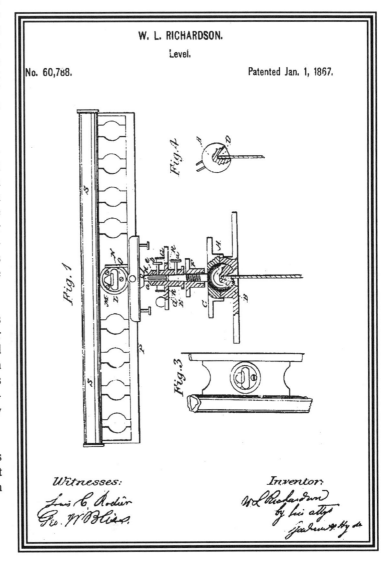

IMPROVEMENT IN COMBINED LEVEL AND PLUMB

In the 1869 Holyoke City Directory, a Patrick Clifford was listed as a laborer at the Hadley Co. (a thread mill). In 1871, Clifford is again listed as employed by Hadley, but no occupation is listed. A second Patrick Clifford (1869) was listed as a blacksmith at Hamden Mills in Holyoke, but there is no subsequent listing. Hadley Corporation had a great machine shop at its mill in Holyoke and eventually did a lot of outside work. It would be a logical maker for Clifford's levels.

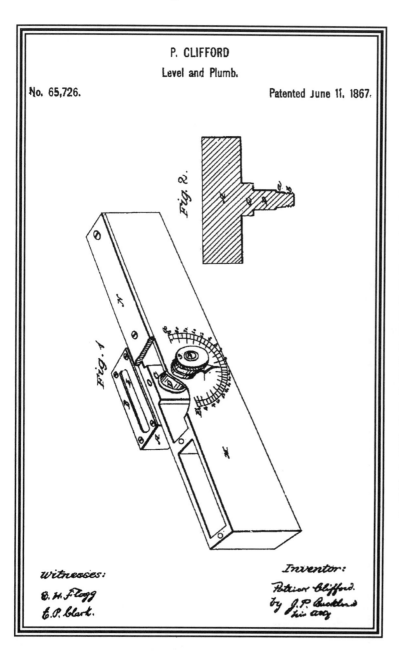

P. CLIFFORD
Level and Plumb.
No. 65,726.
Patented June 11, 1867.

Fig. 2.

Fig. 1

Witnesses:
O. H. Flagg
E. P. Clark.

Inventor:
Patrick Clifford.
by J. P. Buckland
his atty

This patent concerns a vial coffin mounted on the end of a shaft. The shaft protrudes through a hollow frame that is shaped like an ordinary level stock. There is a pointer on the shaft, on the opposite side of the stock, and a scale is placed on the stock on that side. Thus the invention is a manual inclinometer.

This is the patent date on all known Clifford levels although there is, apparently, no level in this specific configuration. All known Clifford levels use the configuration of the later patent (see November 26, 1867). An implication of this practice (using only the early patent) is that Clifford changed his design and then began manufacture of his levels prior to the granting of the second patent. It appears that the iron level (Nov. 26, 1867) was marked with his first patent date to try to protect his invention until the second patent was granted.

IMPROVEMENT IN PLUMB LEVELS

In 1867-68, both Ensminger and Elmer were employed at the US Armory in Springfield, MA. Sometime in late 1868 or early 1869, Elmer left and went to work for the NY Watch Co in Springfield. By 1870-72, Elmer was employed by J. B. Rumrill & Co (manufacturing jewelers). In 1872-73, he was employed by Smith and Wesson, but by 1873-74, he was said to be employed by Newell Bros (button maker). From 1874 to 1876, Elmer was again employed by the U.S. Armory, but 1876-77 saw him back at Newell Bros. Ensminger is not listed further in the Springfield directories.

The patent drawing is obviously mistitled. The inventors thought of the invention as a universal level and protractor. The vial coffin was carried upon a shaft behind the protractor scale. The shaft could rotate and be set at any desired angle. The indicator arm could rotate with the shaft or be independently set as desired. The claim is for the indicator arm and the scale being used as a bevel in conjunction with the rest of the device.

This device could have been inspired by Clifford's first patent (June 11, 1867). It is essentially the same as Clifford's patent except that the iron stand is substituted for the wooden stock.

No such device has been observed to date, although this level is sufficiently simple and apparently useful that manufacture seems probable.

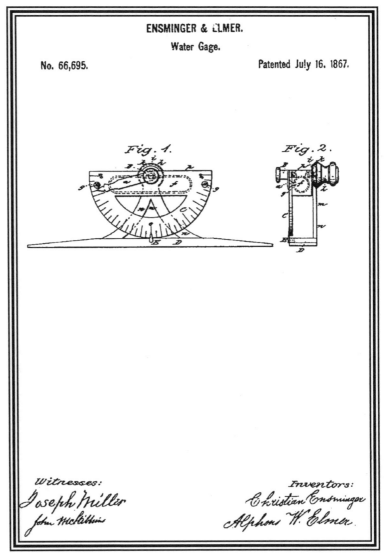

ENSMINGER & ELMER.
Water Gage.
No. 66,695.
Patented July 16. 1867.

Fig. 1.

Fig. 2.

Witnesses:
Joseph Miller
John McLithin

Inventors:
Christian Ensminger
Alphons W. Elmer

IMPROVEMENT IN ADJUSTING SPIRIT LEVELS

Samuel Chapin was listed in the 1870-71 New Britain City Directory as a rule maker. Subsequently, he was listed as mechanic and machinist at the Stanley Rule & Level Co. until 1904, when he moved to Newington.

Augustus Stanley was a founder of A. Stanley & Company.

The top plates of the level depicted in the patent drawing show a vial casket that is an integral part of the top plate. This is much like earlier levels such as the Nº 20 & Nº 25 of Hall & Knapp, except that here the claim was for vulcanized rubber washers under adjusting screws on vial casket instead of springs.

The patent also claimed a plumb fixture that was to incorporate the same type of adjustment. This part of the claim is somewhat vague regarding the physical arrangement.

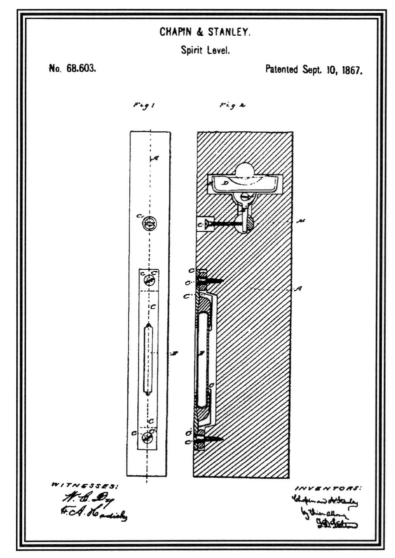

CHAPIN & STANLEY.

Spirit Level.

No. 68.603.

Patented Sept. 10, 1867.

Fig 1   Fig 2

WITNESSES:

INVENTORS:

Stanley levels carrying the Sept. 10, 1867 patent date are relatively common, but they don't incorporate any of the claims of this patent regarding the plumb fixture. Instead, they utilize the plumb adjustment features of Traut's patent of October 6, 1868.

The vial casket/top plate combination with the springs on the adjusting screws had also been in common usage in New York and New Jersey for at least 20 years.

L. L. Davis   September 17, 1867
Springfield, Massachusetts   68,961

IMPROVEMENT IN ADJUSTABLE SPIRIT LEVELS

Leonard Leroy Davis was born in Gilmanton, NH on February 21, 1838.[5] He was considered a native of Laconia, NH and had established his home there and died there on August 13, 1907.[6] Davis was first listed as a manufacturer of spirit levels in the 1868-69 Springfield City Directory. He had started the tool and level business in Chicopee Falls in 1867 and signed the pieces, L. L. Davis. In 1875 he changed the name of the business to Davis L & T Company and moved to 30 Taylor Street in Springfield, MA, a large industrial building that must have housed several small manufacturing companies. (This address is also sometimes given as Dwight and Taylor.) At its peak during the 1880s, the business on Taylor Street in Springfield occupied one floor of a building with dimensions of 60' x 90'. He was said to employ from 15 to 25 "first class mechanics", and to have had considerable export business to Great Britain, France, Germany and Australia. *The Inland Massachusetts Illustrated*[7] described his business as follows:

> "The specialties of this house embrace full lines of hardware specialties and machinists' supplies, including adjustable spirit plumbs, levels, and inclinometers, iron pocket levels, builders' levels and level glasses, patent ratchet bit braces, improved iron-block planes, carpenters' and machinists' marking awls, jewelers' screwdrivers, hack saws, breast drills, thread gauges, drill and lathe chucks, combination calipers, Marshall's universal squares, planer jacks, Buell's "Giant" hollow handle tool sets, "Champion" amateur scroll chucks, independent reversible jaw chucks, and many other useful and valuable implements. A new and important device that will interest all railroad men is Johnson's patent car brake, which operates by means of a lever, lies low, is situated where the brakeman is exposed to no danger, is instantly applied and released, is strong and light, is cheap, durable, and reliable, takes up all slack caused by wear, all parts made in duplicate, saving cost of repairs; the frames are interchangeable, are out of the way of brakeman's lantern, and the long lever down when the brake is on or off."

Davis' interest in railroad-oriented devices, as evidenced by the car brake, should be no surprise. He started his career as a blacksmith's apprentice but soon went to the Randlet Car Company. He then moved to the repair shops of the Boston, Concord & Montreal Railroad, then to the Manchester Locomotive Works, followed by the Hinkley & Williams Locomotive Works, Dearborn, Robinson & Co. iron works, Denio & Roberts, safe manufacturers, Allen & Endicott, manufacturers of steam-engines and general machinery, finally returning to his former position on the Boston, Concord & Montreal Railroad. Yet by the time he left the railroad again he was only 23 years old. He next worked at the E. and T. Fairbanks Company's Scale Works at St. Johnsbury, VT. In late 1862 he went to Springfield to take a position at the U.S. Armory there. In late 1862 or early 1863 he went into business for himself manufacturing a bolt heading machine according to his own patent. Railroads and locomotive works were among the primary customers. By mid-1864 he had sold the business and retired. He was now 26 years old. His next business venture was a traffic in general machinery and railway supplies, and while doing this he obtained the initial level patents. It can be said that Mr. Davis stayed longer at the business of tool manufacture than at any other, but even before he left that business he was making supplies needed by the railroads;

---

[5]   This discussion of Leonard Davis is adapted from Rosebrook, Donald, *American Levels and Their Makers, Vol.I, New England.* Astragal Press, Mendham, NJ 1999, pp 57-59.

[6]   Smith, Roger K., *Patented Transitional & Metallic Planes in America, 1827-1927*, The North Village Publishing Company, Lancaster, MA 1981. pp 180-182.

[7]   *Inland Massachusetts Illustrated: Hampden, Hampshire, Franklyn, and Berkshire Counties*, Springfield Mass, The Elstner Publishing Co. (1890) p 56.

when he did leave the tool and level business he returned to the business of railway supplies,, The Electric Railway Switch and Supply Co., and also pursued a new interest, a venture into the production of incandescent lights.[8]

Davis maintained various other offices during his stay in the Springfield area. In 1870, besides the manufacturing facility in Chicopee Falls, he had an address of 10 Fort Block at 292 Main in Springfield, MA, usually described as offices and rooms. In 1871 he claimed in an ad to having an office over the John Hancock National Bank in Springfield. He maintained the Chicopee Falls address until sometime in 1874. In 1875-76, Davis L & T was said to be located at the corner of Dwight and Taylor. In 1876, Davis L & T was listed in Brightwood, which was said to be two miles north of City Hall on the Connecticut River Railroad. This latter may have been some type of intermediate processing for his casting, or a similar operation.

By 1887-88 his ads began to claim the manufacture of machinists' tools and railway supplies. His 1889 ad in the Springfield City Directory does not even mention levels and his bill head in 1890 likewise did not mention levels other than in the company name. However, levels returned in the ad in the Springfield City Directory of 1890.

In 1891 and 1892, L. L. Davis was listed as the general manager of Davis L & T (at 30 Taylor) and as a manufacturer of incandescent lamps (at 32 Taylor). This is the last listing for Davis Level & Tool to include railway supplies, bit braces, and machinists' tools. In 1893, 30 Taylor is occupied by J & H Duckworth, machinists' and sewing machine parts. In 1894, the directory lists Davis Electrical Works at that address. In the 1895 directory L. L. Davis was listed as Treasurer and General Manager of Davis Electrical Works. In 1900 there was no Davis Co. listing. In 1901 Davis was president and manager of the Electric Railway Switch & Supply Co. located at 30 Taylor St. In the 1905 directory, Davis was shown as Vice President and General Manager of Davis Electric Mfg. Co., located at 30 Taylor St. By 1905, Davis had already returned to Laconia NH for a short two years in retirement before his death. It would appear that Davis never moved his family to Springfield. Various Springfield City Directories list him as boarding at different locations in the area.

This is the basic inclinometer patent upon which Davis founded his success. The claim was for (a) semicircular bubble case; (b) the face ring carrying the graduations; (c) the dove-tailed socket for the face ring and the associated set screws; and (d) an elevating screw at one end of the base with calibrated rise per turn. The original concept used a one foot scale along the base as well as the elevating screw. Neither of these latter two features is known to exist. The proposed mantle clock stock was much fancier than known models, the envisioned scale was only 180° and the side screws were much more prominent. But the lasting concept was clearly proposed here.

---

[8]   *The Leading Citizens of Hampden County: Biographical Review*, Boston, Biographical Review Publishing Co., 1895, pp 943-944.

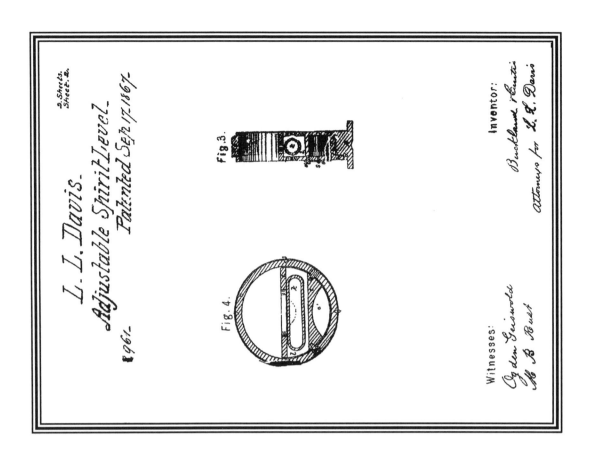

L. L. Davis.
Adjustable Spirit Level.
Patented Sep. 17. 1867.

2 Sheets.
Sheet. 2.

No. 961.

Fig. 3.

Fig. 4.

Witnesses:
Ogden Griswold
Geo. B. Post

Inventor:
Buckland Austin
Attorneys for L. L. Davis

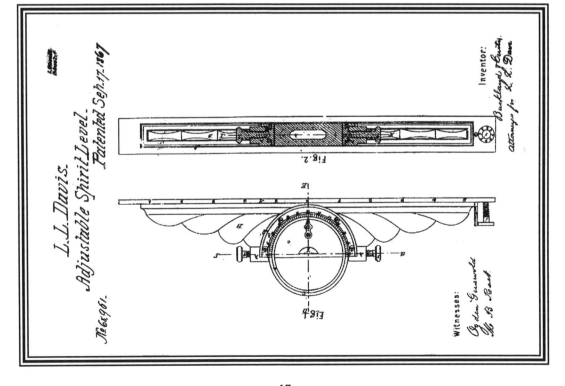

L. L. Davis.
Adjustable Spirit Level.
Patented Sep. 17. 1867.

No. 69,961.

Fig. 2.

Fig. 1.

Witnesses:
Ogden Griswold
Geo. B. Post

Inventor:
Buckland Austin,
Attorneys for L. L. Davis

Patrick Clifford
Holyoke, Massachusetts

November 26, 1867
71,279

Assigned to Himself and James Doyle of Holyoke, MA

IMPROVEMENT IN ADJUSTABLE SPIRIT LEVELS

James Doyle was a grocer (Doyle and Finn) in 1869, with a business located on the corner of High Street and Lyman near the Hadley Corp. In the 1871 Holyoke City Directory, the business of Doyle and Finn was now said to be groceries and furniture. By 1874, Doyle and Finn were said to be involved in a wood and coal yard, furniture, groceries and exchange and immigration services.

For information about Patrick Clifford, see his patent of June 11, 1867 on page 12.

This is Clifford's second patent and, except for the substitution of the iron frame for the wooden stock, the device is broadly similar to that proposed in the first patent. In this case, however, the shaft that protrudes from the level casket has a semi-circular sector attached to it, and that sector rotates when the casket rotates. It is fitted with a stop that acts as a detent at zero inclination. The sector is graduated and toothed around the perimeter, and these teeth engage a protrusion on a leaf spring attached to the base of the stock. The casket cover carries the June 11, 1867 patent date.

This device differs from the inclinometer of Ensminger (patent July 16, 1867) by the addition of the tooth wheel and the spring.

This is the form of the tool that is normally found, in spite of the fact that the tool carries the June 11 patent date.

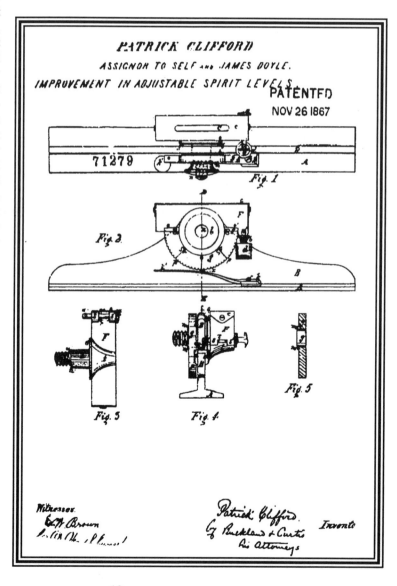

IMPROVEMENT IN PLUMBS AND LEVELS

Seymour Bostwick was a 34 year old machinist, at the time of the patent. He was originally from New York, and was married with two children. He was not listed in either the 1860 or the 1880 census.

Bostwick's patent was for a manual inclinometer. A semicircular block containing the single standard spirit vial is mounted in a matching cutout on a level stock. The mounting plate or side plate contains graduations as on a protractor. A thumb screw through the mounting plate and into a channel in the block serves to secure the block in any desired position.

To date, no example of this patent has been identified.

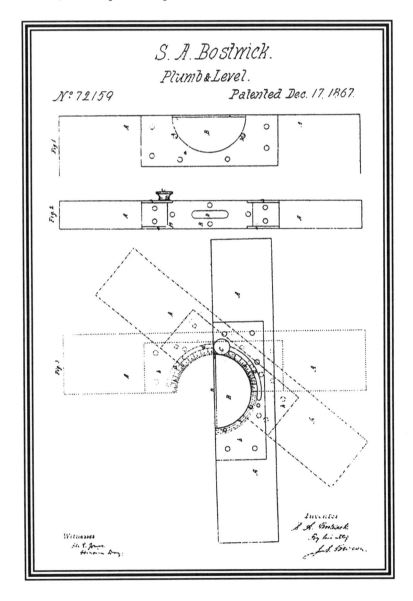

IMPROVEMENT IN ADJUSTABLE SPIRIT LEVELS

William Tate was listed in 1867-68 New Haven City Directory as a machinist. He had left New Haven by the time the 1869-70 Directory was issued. Since his patent was utilized by William Johnson of Newark, NJ, it is possible that he was hired by Johnson and moved to New Jersey.

This patent for an adjustable level utilized two screws through the top plate, with springs between the top plate and the level casket. No claim was made for an adjustment for the plumb vial.

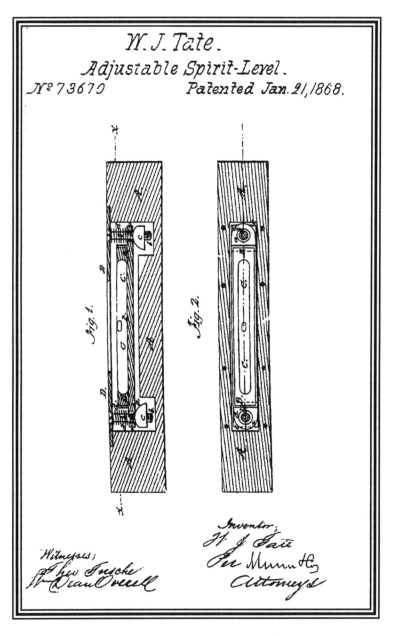

This was the only second patent for an adjustment mechanism on carpenters' levels (the Chapin - Stanley patent preceded this by four months). The use of springs would prove to be a more enduring mechanism than compressible rubber or leather gaskets and the incorporation of springs into subsequent patents by other inventors became widespread.

This patent differs from the single piece top plate/vial casket arrangement by separating the two pieces and placing the springs between them. Another difference is the use of machine screws for connecting the pieces and adjusting the attitude of the casket.

IMPROVEMENT IN SPIRIT LEVELS

For information about Leonard Davis, see his patent of September 17, 1867 on pages 15-16.

This patent, Davis' second, describes the incorporation of a spring and a holding screw to vary pressure on the rotating ring. In addition, the patent describes the adjustments for the detents at 0° and 90°. The adjustments utilize screws with conical points that interact with beveled studs protruding from the rotating sector. Here, the drawing implies a standard metal level stock with the holding screw in the top rail.

All of the aspects of this patent are found utilized, and marked, only on the tall frame inclinometers. However, many other inclinometers, incorporating the spring and holding screw, are known; and, in fact, nearly all mantel clock inclinometers, such as proposed in Davis' first patent, have the provision for the spring.

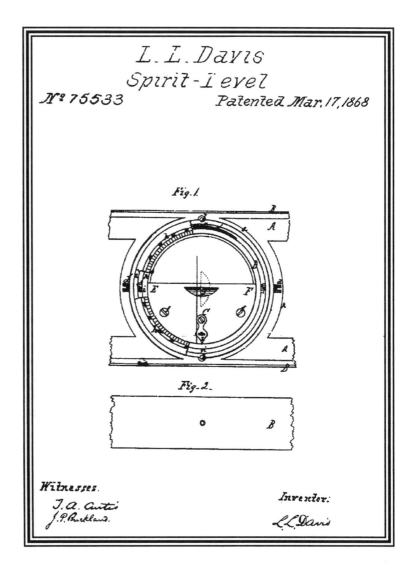

IMPROVEMENT IN ADJUSTABLE SPIRIT LEVELS

For information about Leonard Davis, see his patent of September 17, 1867 on pages 15-16.

The patent entails a pin that acts as a pivot for the level casket. This pin goes through a hole in a fin at the bottom of the casket. Adjustment of the position of the casket is accomplished by turning screws on either end of the casket. The screws protrude through the top of the top plate and when they are turned in opposite directions the casket rocks on the pivot pin.

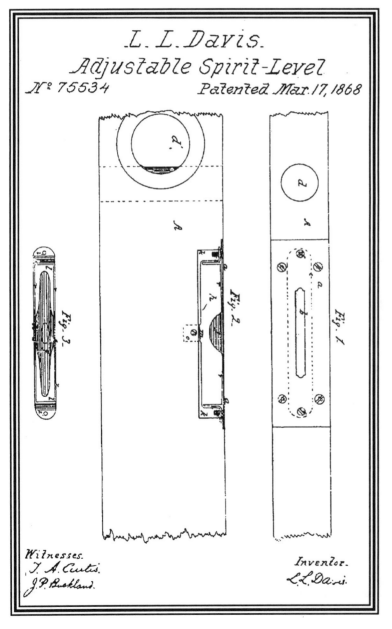

This is the patent employed in the early Davis adjustable wooden carpenters' levels. While he continued to use brass marked with a patent date of March 17, 1867 (a mistake), Davis changed the adjustment mechanism at some point during his manufacture of wooden levels.

IMPROVEMENT IN SPIRIT LEVELS

Sibley was a Middletown, CT architect and building superintendent.

The claim was for a revolving, sighted level in combination with the circular disk. A standard wooden carpenters' or masons' level was to be modified by the permanent addition of metallic plates *D* on each end of the top. The plates were to be grooved to receive sighting pieces that could be reproducibly put into place when needed, and removed when not in use. One sight had a peep hole and the other had crossed wires. A socket was envisioned for the center of the bottom of the level, and that socket was to fit onto a pivot point in the center of the table.

The Sibley's patent sighting level, as manufactured, deviates substantially from the description in the patent. The device described in the patent had a thick circular base constructed of wood or metal, with degrees marked around the circumference. A modified wooden stock level was specified in the patent.

In actuality, the level is made of cast iron and brass and was constructed specifically for the table. The table is constructed of cast iron and has the degrees laid out on the upper edge of the circumference.

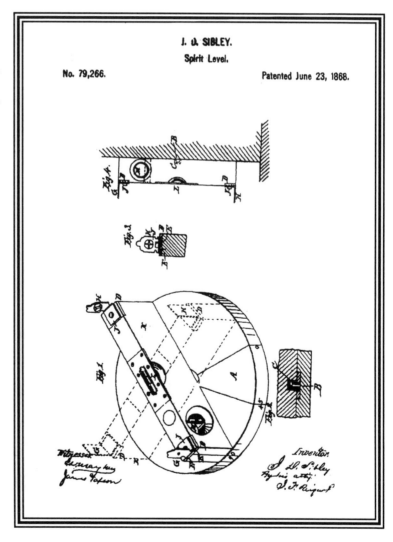

IMPROVEMENT IN SPIRIT-LEVELS

Homer Lewis died in the Brattleboro Insane Asylum in 1884. His obituary[9] claimed that 12 years previous he "was worth some property", had a large circle of friends and was a genial, well educated and popular fellow. "He became dissipated, lost his money, and finally becoming insane, he was sent by the town authorities to the State institution where it was soon found out that his case was incurable." The obituary claimed that nothing else was known about him. He was not counted in either the 1860 or the 1870 census.

This patent claimed a simple and apparently practical vial adjustment mechanism. The mechanism used a u-shaped leaf spring at one end of the level vial casket and another in the center of the plumb casket. Adjusting a screw would compress or loosen the appropriate spring.

To date, no examples of this device have been identified.

H. LEWIS.
Spirit Level.

No. 79,363.

Patented June 30, 1868.

Witnesses;

Inventor;
H. Lewis

---

[9]   The source of the obituary is missing and unknown at this writing.

IMPROVEMENT IN LEVELS

Hiram Loomis had been an employee of Hermon Chapin in the rule making department[10] In 1867-68, he was found to be a civil engineer and surveyor in Hartford and he had apparently been in that business since 1860, or perhaps before. He became Hartford City Surveyor in 1870.

The patent envisioned a single hinged sight *A* and the second sight *B* on a calibrated slide. The slide was to be calibrated in feet per hundred feet. Sighting was to be through the slits.

The product manufactured by H. Chapin's Son (marked H. Chapin) was faithful to the patent obtained by Loomis. This patent was probably inspired by Sibley's patent and followed it by only two weeks.

The actual product does not have calibration on the slide, but the owner could presumably do the calibration himself.

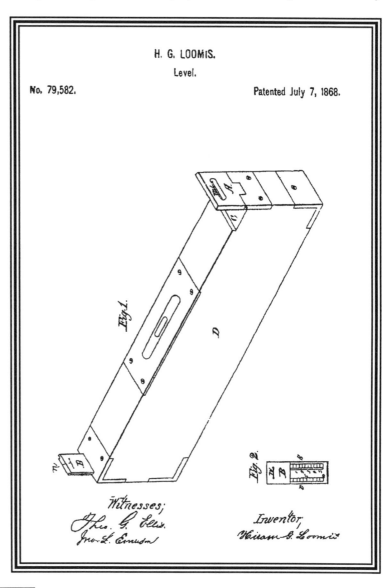

---

10   Roberts, K. D., Op. cit., page 113

IMPROVEMENT IN LEVELS

William P. Cutter was born in Chelsea in 1834 according to the Grave Records of Woodlawn Cemetery. In 1866 and again in1868, Cutter was listed in the Chelsea City Directory as a ship's carpenter with a house at 34 Walnut Street. In 1868 through 1874, he was listed only as a carpenter with a house on Cottage Street.

The device consists of a through-bored stock provided with a metal ring on each side of the bored hole. The intersection of cross pieces at right angles to each other, in each ring, serve as a bearing point for a shaft carrying the pendulum. There were no graduations of any type so, essentially, the device could show plumb and level only.

The device is a pendulum type plumb and level, otherwise called an inclinometer. No inclinometers corresponding to this particular patent have yet been observed.

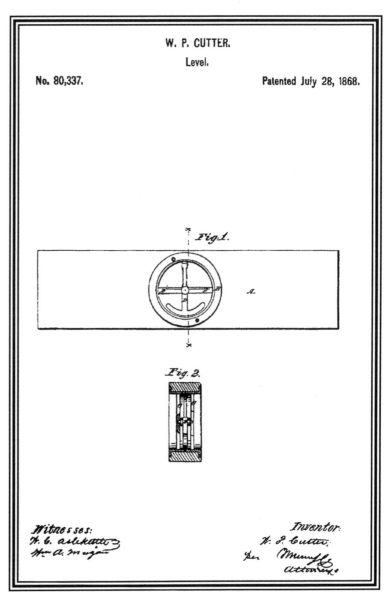

W. P. CUTTER.

Level.

No. 80,337.   Patented July 28, 1868.

Fig.1.

Fig. 2.

Witnesses: Inventor:

DESIGN FOR A SPIRIT LEVEL

For information on Leonard Davis, see his patent of September 17, 1867 on pages 15-16.

Davis describes the outer frame (circular) of the level and the bed pieces (rails) and addition frame pieces (end posts) "...and the space which this frame encloses is filled up or occupied by the scroll or ornamental work..."

This is the design patent covering the inclinometer in its rectangular metal stock form with the filligree infill.

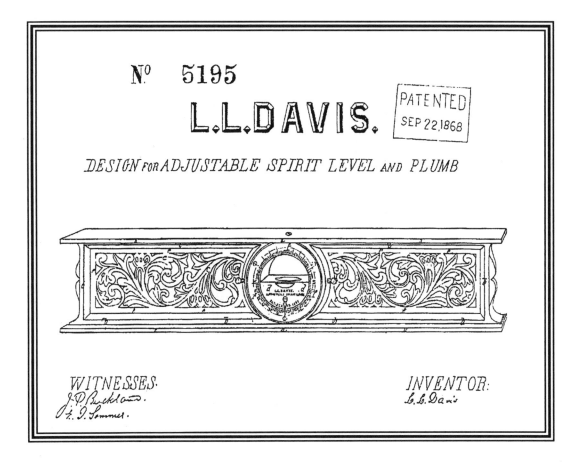

Assigned to the Stanley Rule and Level Company of New Britain, CT

IMPROVEMENT IN SPIRIT LEVELS

Justus Traut was born in Pottsdam, Germany in 1839. He followed his father to America in 1854 and they both went to work for Hall & Knapp. Upon the merger of Hall & Knapp and A. Stanley & Co. in 1858, both Trauts became Stanley Rule & Level Co. employees.[11] Justus Traut was granted at least 28 patents pertaining to levels. As will be seen, he had patents for adjustments, independent sights, and level stocks; but his most enduring patents involved the machinery for making lines on vial and the venerable hand-y grip.

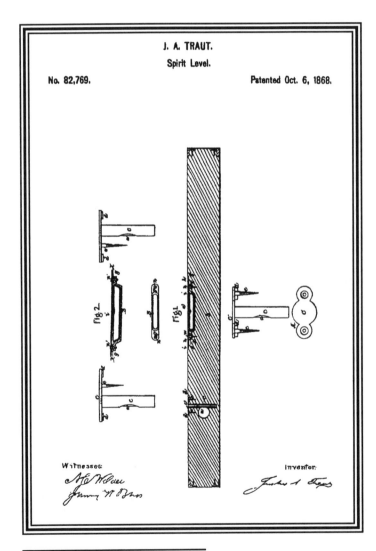

This patent describes a means of adjusting both the plumb and the level independently. The patent allows the use of springs, as well as compressible rubber grommets underneath the adjustment screws.

The application of this patent was very faithful to the idea described in the patent. This patent was incorporated in Stanley levels, as were a large proportion of Traut's other patents applicable to levels and leveling devices. The plumb assembly will be familiar to persons who have examined Stanley levels.

The style shown in *Fig. 1* was used in Type 2A levels, and the style shown in *Fig. 2*, was used in *some* Type 3 levels and, with a change in the top plate design, in subsequent levels. The vial casket used in Type 2A levels does have cutout sides in the center as suggested by the patent.

Type 2A Stanley levels appear to incorporate all of the features of this patent rather than the Chapin-Stanley patent.

11　　Smith, R. K., *Patented Transitional & Metallic Planes in America, Vol. II*, R. K. Smith Publishing, Athol, MA 1992, p 207. Additional information is available in this reference.

IMPROVEMENT IN SPIRIT LEVEL

Andrew P. Odholm was listed in the 1868-69 Bridgeport City Directory as a carriage maker living at 150 Broad St., and in the 1869-70 directory as an ornament maker at Simmons Dock. In 1872-73, he was listed as a stair builder living on N. Bridgeport. In prior years, he had sometimes claimed to be a wheel maker, which is consistent with the carriage maker trade, while other times he is found under the more general occupation as carriage maker. He was not listed prior to 1857 and is known to have died prior to the publication of the 1876 city directory.

The patent is for a level consisting of a liquid-filled circular chamber sealed to glass with vulcanized rubber. The claim was that prior instruments using this type of circular chamber were constructed of metal, which sprung leaks due to expansion and contraction of the metal. His invention, however, utilized the rubber being shrunk onto the glass and onto the dial (glass or metal) in the rear. The chamber was then filled through a hole in the rim.

The concept of this patent appears to be based on Vandergrift's patent of May 8, 1866; a tool of Vandergrift's design was manufactured in Bridgeport by the Patent Level Company. However, no example of this device has yet been identified.

Note the title of the drawing!!

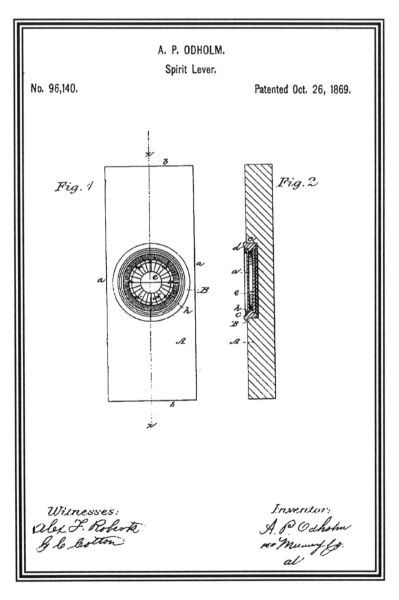

A. P. ODHOLM.

Spirit Lever.

No. 96,140.

Patented Oct. 26, 1869.

Fig. 1

Fig. 2

Witnesses:

Inventor:

Edwin A. Stratton and Charles M. Stratton
Greenfield, Massachusetts

March 1, 1870
100,463

IMPROVEMENT IN SPIRIT LEVELS

According to the Gravestone records of Green River Cemetery, Edwin Stratton was born September 15, 1819, and Charles Stratton was born July 19, 1823. Edwin and Charles, together, worked as house carpenters until 1862 - 1865, during which time Edwin was employed in the Springfield Armory. They continued to work as carpenters after the war, until they established their level business. The brothers worked together until Charles died in 1893. Edwin sold the company to his son-in-law, Raymond Stetson, on February 15, 1902. Edwin died in 1906.

The patent describes the use of grooves at right angles to one another to seat the rods that formed the brass binding. Note that the rods do not project through the end plates but that end plates are suggested to prevent the rods from coming out.

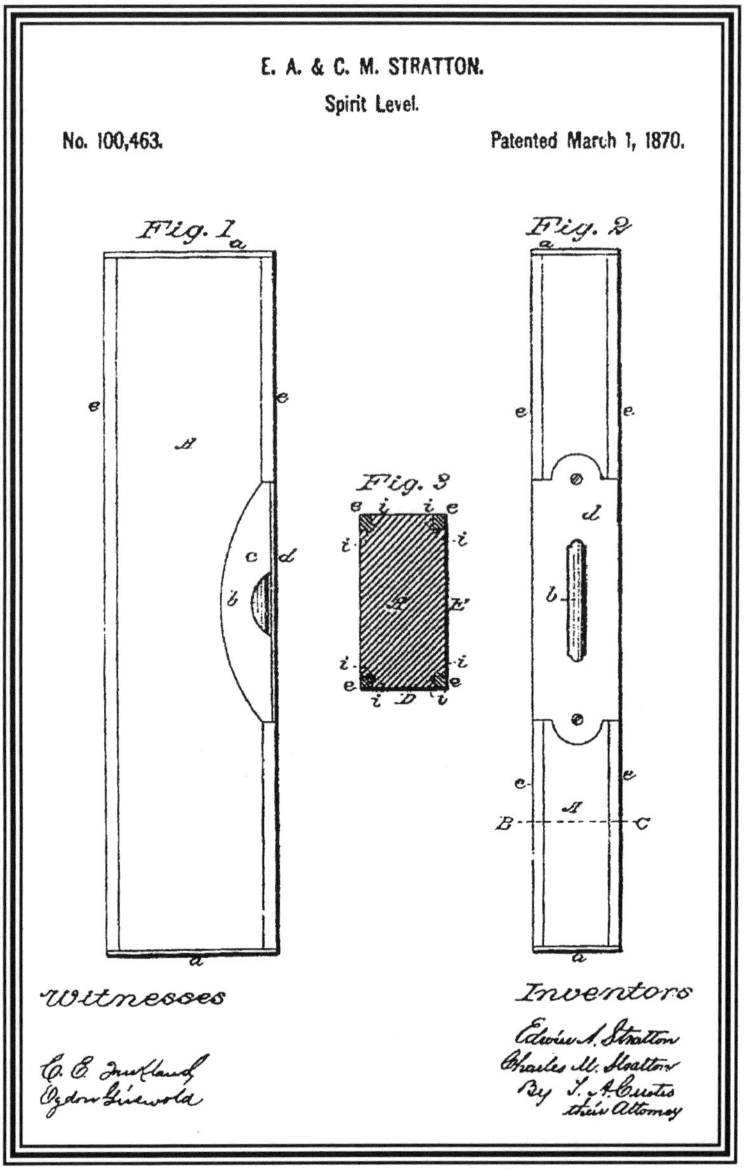

The patent suggests using a small circular saw to make cuts at the edges at right angles to one another, such that a small square piece is removed and slots are left to hold the ribs on the brass rods.

This, the Strattons' first patent, was for the rods and the process to bind the edges of wooden levels. It was used on all brass bound Stratton levels for over 40 years. After the patents expired, it was used by many other makers such as the Stanley Rule & Level Co. and Chapin-Stephens for many years.

E. A. & C. M. STRATTON.
Spirit Level.
No. 100,463.
Patented March 1, 1870.

Fig. 1   Fig. 2   Fig. 3

Witnesses
C. C. Buckland
Ogden Griswold

Inventors
Edwin A. Stratton
Charles M. Stratton
By S. A. Curtis
their Attorney

IMPROVEMENT IN COMBINED SQUARE, PROTRACTOR, LEVEL, & c.

No information was located concerning J. A. Littlefield.

The proposed device appears to be a T-square with a spirit vial in the handle in the normal manner. The tongue is adjustable in terms of its angle with the handle, and has both standard ruling marks and a protractor. The drawing below is self-explanatory.

To date, no examples of this device have been located.

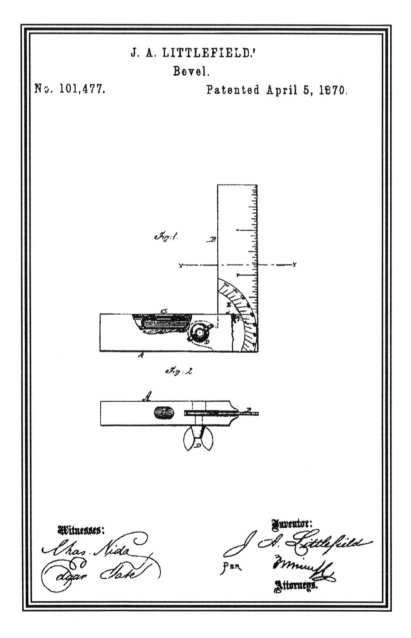

J. A. LITTLEFIELD.'
Bevel.

№. 101,477.   Patented April 5, 1870.

Fig. 1.

Fig. 2.

Witnesses: Inventor:
Chas. Nida   J. A. Littlefield
Edgar Tate   Per   Mun...
                              Attorneys.

IMPROVEMENT IN SPIRIT LEVELS

There was a George A Shelly, who was a partner in "Shelly and Treat" (stoves and tinware), in 1872-73 in New Haven[12]. During the period from 1876 to 1884, the New Haven City Directory listed Shelly as a tinner at J. S. Rice in New Haven. He was listed in the directory with no occupation in 1890 and was not listed after 1893. George Shelly also has other patents, including one for a different type of bevel square. That patent was entitled "Improvement in Square and Bevel Combined" and was issued on April 5, 1864. The patent was utilized by Stanley Rule & Level Co. in producing their № 24 "Combined Trisquare and Bevel", beginning in 1867.

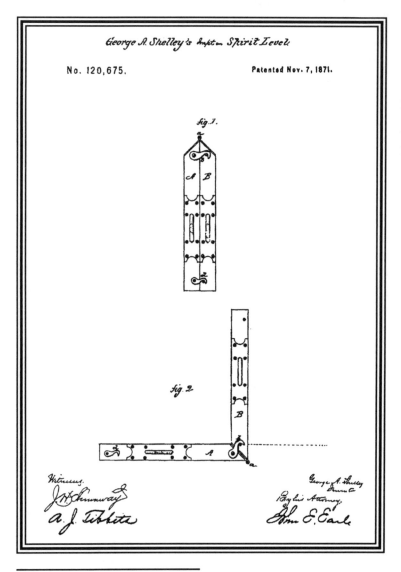

This patent is for a hinged spirit level that can be locked closed, or can be opened and locked into a square. It was envisioned to contain two spirit vials and thus, in the open position, it became a kind of cross-test level.

No examples of this tool have yet been identified.

IMPROVEMENT IN ADJUSTABLE SPIRIT LEVELS AND PLUMBS

For information on Leonard Davis, see his patent of September 17, 1867 on pages 15-16.

As can be seen in the drawing, the plumb vial case has a stud, or "journal," protruding from the top. The screw top, or "plug", has a hole that is off center, or "eccentric," and into which is inserted the stud from the plumb vial case. Thus, by turning the screw top, the plumb vial case and its spirit vial can be moved back and forth.

This is the familiar plumb adjustment mechanism in used in fully adjustable Davis wooden carpenters levels. It merely uses a screw located eccentrically in the top of the plumb vial case and can be adjusted from the top without removing the cover.

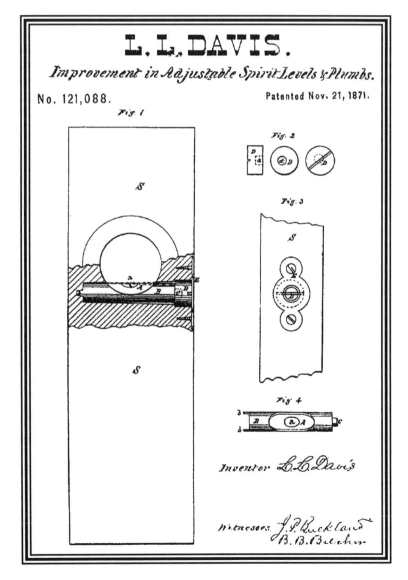

Justus A. Traut
New Britain, Connecticut

July 2, 1872
128,513

Assigned to the Stanley Rule and Level Company of New Britain, CT

IMPROVEMENT IN SPIRIT LEVELS

For information on Justus A. Traut, see his patent of October 6, 1868 on page 28.

The principal of this patent applies equally well to both the plumb & level fixtures. It describes a level casket hanging from the top plate. It is supported with two mounting screws through the top plate, with a spring on one of the screws between the top plate and the level casket.

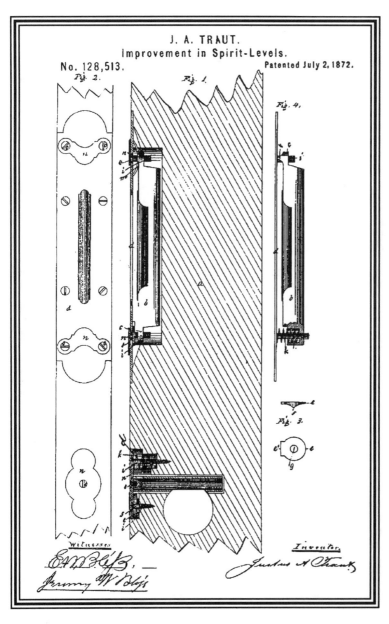

J. A. TRAUT.
Improvement in Spirit-Levels.
No. 128,513.
Patented July 2, 1872.

This patent was applied to the level assembly of the Type 3 levels, and to the plumb assembly of *some* of the Type 3 levels. It may be that Stanley was using up old Type 2A plumb fixtures in the lesser quality levels, but used this new plumb adjustment in higher quality levels Type 3 levels such as the Nº 11. However, only one screw was used in actual practice, and the other end of the vial casket was hung on a clip soldered to the top plate. Other than this departure, the patent drawing in *Fig. 4* is an accurate representation of the adjustment as Stanley applied it. The protective covers for the top of the adjusting screws are accurately represented.

Edwin A. Stratton and Charles M. Stratton
Greenfield, Massachusetts

July 16, 1872
129,183

IMPROVEMENT IN ADJUSTABLE SPIRIT LEVELS

For information on Edwin A and Charles Stratton, see their patent of March 1, 1870 on page 30.

One of the more interesting aspects of this patent is that the spirit vial itself is not adjustable. (No Stratton level ever had an adjustable spirit vial.) What is adjustable is the point on the spirit vial that will indicate a level condition. The device described in the patent is a bar held by a screw through a slot at one end. Adjustment requires removing the top plate and loosening the screw. The bar, and its indexing point, could then be moved and the holding screw tightened again.

This patent was apparently not used in the form described. However, on levels with this patent date, the moveable wire adjustment is found. It still requires removing top plate and loosening the screw that holds a small plate over the ends of the wire loop. The wire, bent in the shape of a U with long legs, is then adjusted to indicate a level condition, and the small plate over the ends of the wire is tightened again.

This is the first of the Stratton patents for an adjustable indicating or indexing device. This patent was employed on the level fixture of all of their adjustable levels until it was replaced by the unpatented little brass bridging strip that fit between the channels of the brass binding strips.

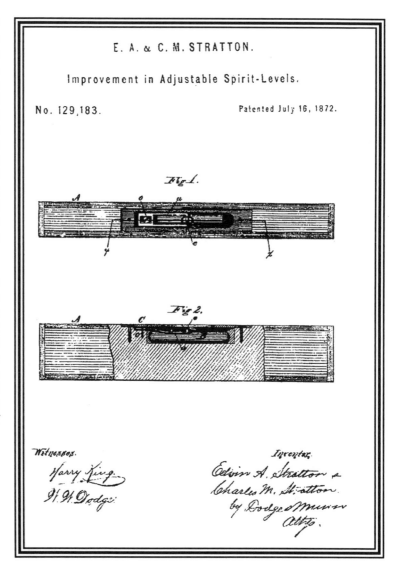

E. A. & C. M. STRATTON.

Improvement in Adjustable Spirit-Levels.

No. 129,183.   Patented July 16, 1872.

Fig 1.

Fig 2.

Witnesses.
Harry King
H. H. Dodge

Inventor.
Edwin A. Stratton &
Charles M. Stratton.
by Dodge & Munn
Atty.

IMPROVEMENT IN PLUMBS AND LEVELS

Sanderson was a "second hand" at Wamsutta Mill in New Bedford during the period immediately before and after the date on this patent. He had a house there, on Hazard St.

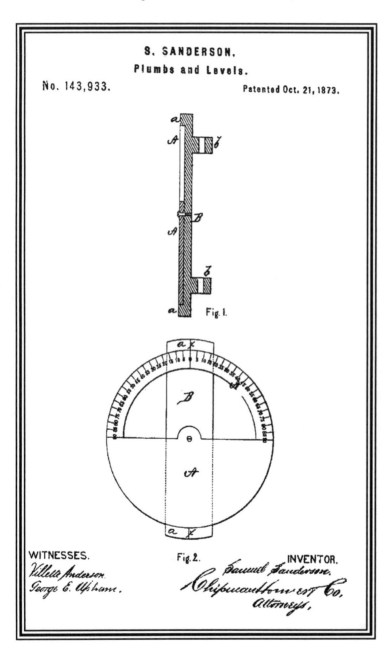

S. SANDERSON.

Plumbs and Levels.

No. 143,933.

Patented Oct. 21, 1873.

Fig. 1.

Fig. 2.

WITNESSES.
Villette Anderson.
George E. Upham.

INVENTOR.
Samuel Sanderson.
Chipman Hosmer & Co.
attorneys,

This device was not envisioned as a carpenters level. It is an inclinometer of the weighted wheel variety. The wheel was weighted by removing all metal except for a rim on one half. The rim portion is graduated to 180 degrees. The bar portion serves as support for the pivot as well as providing an index mark for reading the graduated wheel. The device was intended for use on the mule spindle in a cotton mill.

The bar is fixed on the spindle and when the angle of the spindle changes, the bar changes also and the moveable wheel rotates to indicate the angle.

Since this was not a type of level normally encountered by machinists or construction trades craftsmen, it may have been produced, but examples are unknown to collectors or students of leveling devices. It is however, a classic type of inclinometer.

IMPROVEMENT IN SPIRIT LEVELS

Albert Hyde was a clerk and collector for the Boston & Albany Railroad (B&A R.R.) in 1876.

The general shape and truss properties of the stock are the primary features of this patent. As such, the patent seems more like a design patent than a device patent. The patentee also specifies a single, long plumb vial that is claimed to give better accuracy than shorter vials. He claims that, compared to a wooden level, his device is lighter (compared to a cast iron bar), stronger, resistant to changes due to moisture, and more ornamental.

Levels have been made according to this patent and conform to the specifications of the patent.

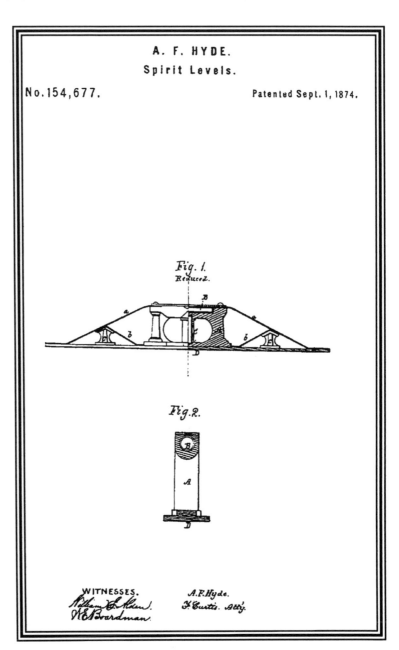

A. F. HYDE.
Spirit Levels.

No. 154,677.

Patented Sept. 1, 1874.

Fig. 1.
Reduced.

Fig. 2.

WITNESSES.

A. F. Hyde.
F. Curtis. Atty.

IMPROVEMENT IN SPIRIT LEVELS

Edwin L. Barnes was a mason in Boston with a business at 76 Church Street, and is so listed in the Boston City Directories from 1869 to 1877. He is said to have moved to St. John in 1878.

The patent calls for the use of a mirror placed so as to reflect the bubble in the spirit vial when viewed from below. The mirror arrangement may be used in either the plumb or level function; however, the level fixture would have to be set in the center of the side of the stock in order to be associated with a mirror.

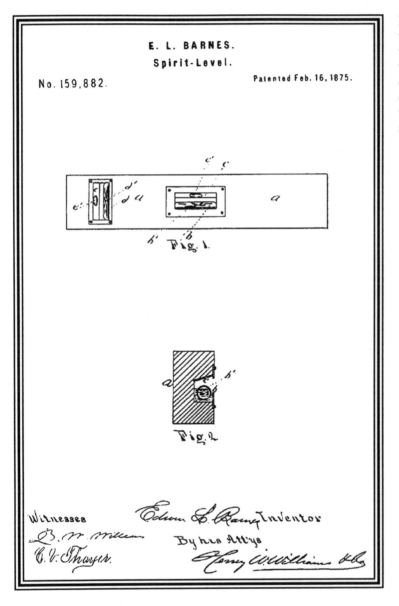

Levels made according to this patent are known, and they appear to be faithful to the patent specifications and drawing. In the observed examples, the level vial is in the center of the side of the stock.

Sylvanus Holbrook
Rockport, Massachusetts

October 11, 1875
168,567

IMPROVEMENT IN RAILWAY-TRACK GAGES

Holbrook was apparently a laborer working during the construction of the railroad into Rockport, and after that time he moved on.

This invention claims a combination of features: (1) a track level with both a permanently fixed bar and a pronged gage at one end, with the pronged gage to be moved forward for curving track; (2) a retractable grade setting bar, with appropriate graduations, at the end opposite the pronged gage; and, (3) a standard level compartment in the top center of the main beam. The retractable gage, which also provides the bar on that end of the device, can be moved up and pivoted into storage in the top of the device. For straight track, the pronged gage is retractable toward the center of the device, exposing the fixed bar on that end.

The invention was manufactured and marketed using his name, and the tool is faithful to the patent. A fine example exists at the Rockport Historical Society Museum.

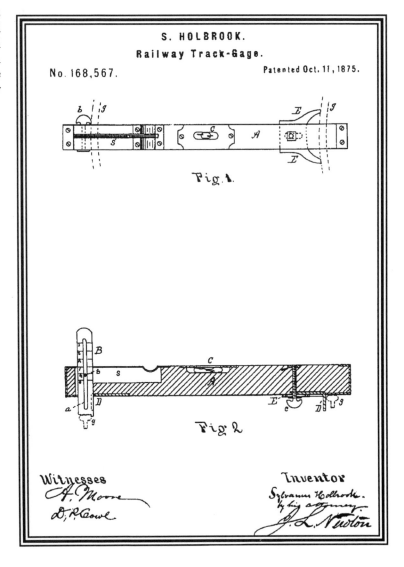

IMPROVEMENT IN SPIRIT-LEVELS

Martin Wilcox was an employee of H. Chapin's Son. He had previously attempted, at least twice over the preceding six or seven years, to patent an adjustment mechanism but was unsuccessful.

Although it is hard to see in the drawings, the patent describes an adjustment that uses two wedge-shaped pieces, one in each end of the mortise. These wedges have elongated holes for the screws that pass through them from the top plate. The vial casket and the top of the plumb assembly have sloping ends. These fixtures are thus adjusted by moving the wedges forward or back.

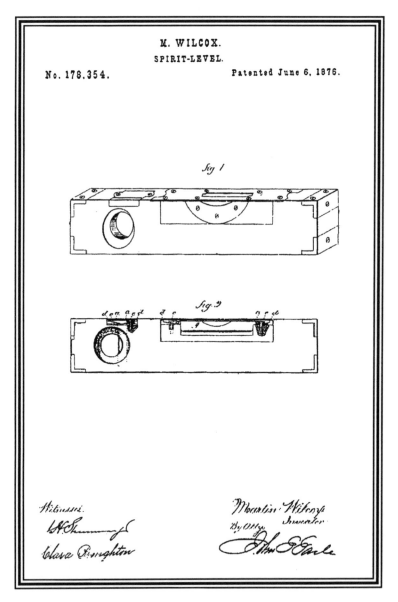

M. WILCOX.
SPIRIT-LEVEL.
No. 178,354.   Patented June 6, 1876.

Fig 1

Fig. 2

Witnesses.
W. F. _____
Clara Broughton

Martin Wilcox
Inventor
By Atty
John E. Earle

This is the adjustment patent used by H. Chapin's Son (and other names for the business operated by E. M. Chapin) and are the only adjustments ever used by the company.

Burkner F. Burlington and Lewis H. Priest
South Lancaster, Massachusetts

182,801

Assigned 100% to Lewis H. Priest

IMPROVEMENT IN SPIRIT-LEVELS

Lewis Priest moved from Lancaster to South Lancaster in 1864. He was listed (w/o an occupation) as living there until 1883. In 1885-86, there is a listing for Mary, his widow, with a house on N. Main.

No information about Burkner F. Burlington has been located.

The patent contains two distinct claims. The first is to provide the vial with an elastic sleeve containing an appropriate opening to view the bubble. The second claim specifies a tube with one or two arms bent at right angles to the main tube. In this way a tube may act as both plumb and level or double plumb if desired.

To date, no such device has been observed.

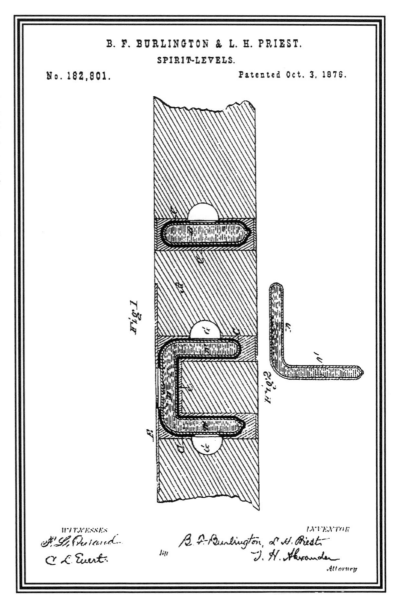

B. F. BURLINGTON & L. H. PRIEST.
SPIRIT-LEVELS.
No. 182,801.   Patented Oct. 3, 1876.

Fig. 1

Fig. 2

WITNESSES

INVENTOR
B. F. Burlington, L. H. Priest
J. H. Alexander
Attorney

DESIGN FOR SPIRIT-LEVELS

For information on Leonard Davis, see his patent of September 17, 1867 on pages 15-16.

The patent description was as follows. "...a frame composed of two plates, ... made flat and true on their outer faces, with an ornamental molding on their inner and opposing faces, with a ring, ... at the center, connecting the two plates, ... and within which is placed a moveable ring,...supporting the spirit tube. Near their ends the plates ... are connected by ornamental rounds ..., and from each of these there extends to the central ring ... a similar round ... arranged horizontally..."

DESIGN.

L. L. DAVIS.
SPIRIT-LEVEL.

No. 10,014.   Patented May 29, 1877.

Witnesses.
Donn F. Tintehell
Will. H. Dodge

Leonard L. Davis.
By his attys
Dodge & Son

This design drawing was submitted as a photograph of a model. The image below is from a copy of a copy of a copy of a 125 year old photograph.[1] The photograph in the copy had darkened considerably and no detail of the design in the iron work could be observed, except for the major features.

The interesting feature of this design, apart from the spindles, is the use of a moveable ring to contain the spirit vial. No such frame is known.

---

[1]   The author is indebted to Mr. Jim Mau for the copies of the photographs in this and the following two design patents. Because the detail has been lost, the author made no attempt to perform an exacting cleanup of the images.

DESIGN FOR SPIRIT-LEVELS

For information on Leonard Davis, see his patent of September 17, 1867 on pages 15-16.

The patent description was as follows. "... design consists of a top and bottom plate, ... made flat and true on their edges and exterior faces, with any suitable style molding or ornamentation on their inner and opposing faces. These two plates ... are connected by a series of rungs or rounds, ... and centrally or midway of the length of the plates is located a ring, ... within which is located the moveable spirit tube."

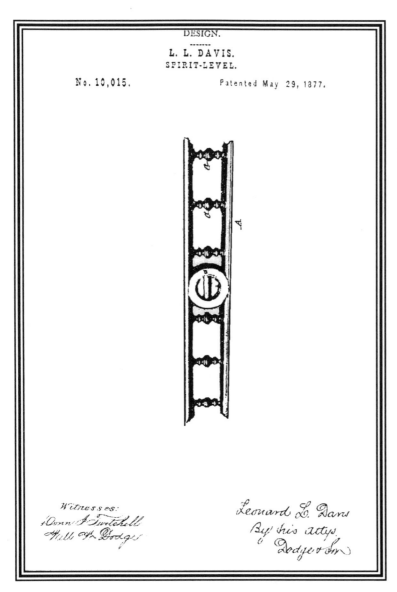

DESIGN.

L. L. DAVIS.
SPIRIT-LEVEL.

No. 10,015.   Patented May 29, 1877.

Witnesses:
Donn A Twitchell
Hall A Dodge

Leonard L. Davis
By his attys,
Dodge & Son

This design drawing was submitted as a photograph of a model. The image below is from a copy of a copy of a copy of a 125 year old photograph. As is the case for the previous design, the photograph in the copy had darkened considerably, and no detail of the design in the iron work could be observed, except for the major features.

As in the case of the previous design, this one calls for a spindle-type frame with a simple inclinometer piece. In this one the inclinometer is located between several vertical spindles connecting the rails. No such frame is known.

DESIGN FOR SPIRIT-LEVELS

For information on Leonard Davis, see his patent of September 17, 1867 on pages 15-16.

The patent description was as follows: "The design consists of two plates, ... made flat and true on their exterior faces and edges, and having their inner and opposing faces either beveled or molded, as may be preferred ... . These two plates, ... are connected by four ornamental cross-bars or rungs, ... and at the center is arranged, parallel with the plates ..., the spirit-tube ..., which is attached to the bottom plate ... by a rest or support, ... which may be made ... in and ... style desired. Near each end a spirit-tube, ... is arranged, in line with the posts or rungs ... and at right angles to the central tube ... . It is obvious that the particular style or form of the rungs ... may be varied at will, and that their number may be more or less, according to the length of the device ..."

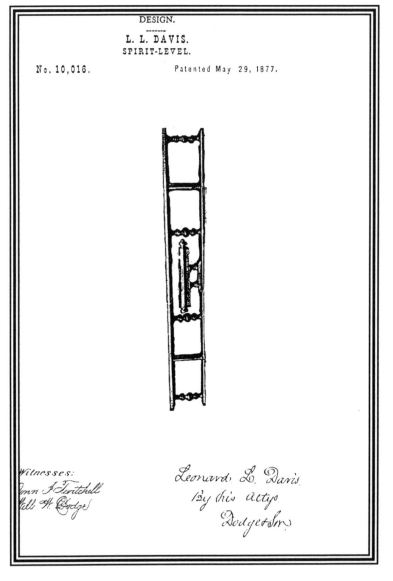

DESIGN.

L. L. DAVIS.
SPIRIT-LEVEL.

No. 10,016.   Patented May 29, 1877.

Witnesses:

Leonard L. Davis
By his attys
Dodge & Son

This design drawing was submitted as a photograph of a model. The image below is from a copy of a copy of a copy of a 125 year old photograph. The photograph in the copy had darkened considerably and no detail of the design in the iron work could be observed, except for the major features.

The design is for a spindle type frame similar to that found in the 6" and 12" iron carpenter's levels. It utilizes vertical spindles only. The known spindle form levels also incorporate another later device patent.

L. L. Davis
Springfield, Massachusetts

May 29, 1877
10,017 Design Patent

DESIGN FOR SPIRIT-LEVELS

For information on Leonard Davis, see his patent of September 17, 1867 on pages 15-16.

The patent description was as follows: "The design consists a base-plate ... made flat and true on its bottom and edges, and having its upper face ornamented with moldings... Upon this base plate ... is mounted a metal tube or case ... which contains the glass spirit-tube ordinarily used in levels, the case ... being supported on a couple of short columns ... I also usually ornament the case ... by making thereon ... ornamental knobs or terminals ..."

This design drawing was submitted as a photograph of a model and no copy of that photograph is available. It is anticipated that the photograph would show a tool similar to the horizontal vial section in the previous design. The design is for a cast iron pedestal level with the level tube mounted on "a couple of short columns." This is the design of Davis' pedestal levels and it was apparently not used for several years after the granting of the patent. These levels also incorporated an adjustment device patented in 1883.

IMPROVEMENT IN LEVELS

Today we would call Oliver Pickering a developer and contractor. He sold land and built houses and was the contractor for several public buildings in Needham.

Pickering's patent describes sights mounted in brass tubes that pop up out of the top of the level stock. On one end the tube contains a pinhole and on the other end there are cross wires in the tube. The sights are propelled out of the stock by springs. The sights are held in the stock by a weighted pawl which drops away when the spring is depressed while holding the level stock at a slight angle.

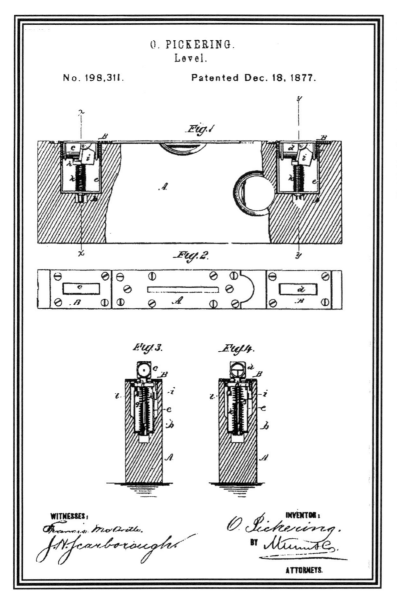

These sights are known, their construction is faithful to the patent specifications, and they look exactly like the drawing below. Pickering suggests that the device is useful for running lines where only approximate accuracy is required. That suggestion shows more realism than that possessed by most inventors of sighting level devices.

TRY-SQUARE

Laroy Sunderland Starrett (1836-1922) was born in China, Maine, on April 25, 1836 and moved to Newburyport, MA, in 1853. He was granted his first patent on May 23, 1865 (for a meat chopper which he began to produce and sell). In 1868, he moved to Athol, where the Athol Machine Co. was to take over the production and sales of the chopper. Athol Machine Company was formed on June 3, 1868 especially to make Mr. Starrett's invention. Starrett was a director but not an officer. For a chronology of his patents and the establishment of his business, (see Cope, K. L., *Makers of American Machinist's Tools*.) Starrett actually invented the combination square in 1876 and, unhappy with Athol Machine Company, he began to make them on his own at Richardson's shop in Athol in 1877. He was forced out of Athol Machine in 1878 and received his patent in 1879, as shown above. Various lawsuits ensued, and by late 1880 or early 1881 L.S. Starrett Co. was created.

This is the classic combination square that we all know and love. The stock of the tool contains a spirit vial and is certainly capable of functioning as a machinist's level.

The manufactured products are faithful to the patent drawings.

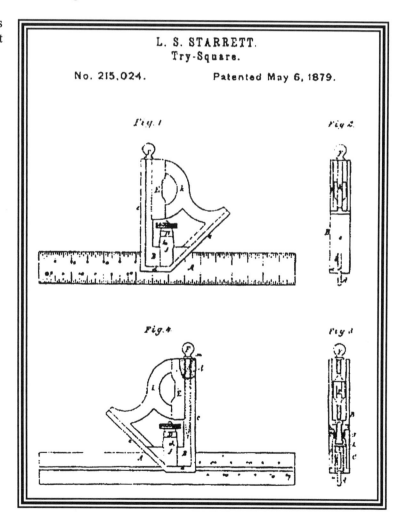

Ira Humphrey
Boston, Massachusetts

May 27, 1879
215,749

Assigned one-half to John L. Gardiner of Boston

IMPROVEMENT IN SURVEYING INSTRUMENTS

No information is available on either the inventor or the assignee. Neither was listed recognizably in the Boston City Directories of the period.

This patent concerns what shall be termed builders levels (no optics). A compass carrying a needle card is suspended from the frame with a binnacle joint. The sight tube has peep hole and cross hairs at the ends. In addition, the sight tube contains a wire *SS'* to assist the viewer in transferring the line of sight to the top of the needle card.. This wire is observable through slots in the top and bottom of the sight tube. The compass also has an arresting spring actuated by a screw on the side of the box.

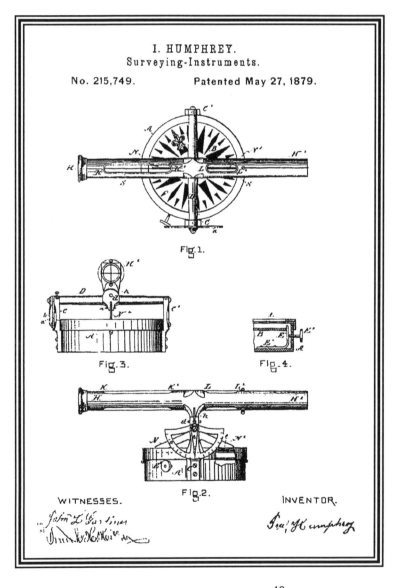

I. HUMPHREY.
Surveying-Instruments.

No. 215,749.   Patented May 27, 1879.

Fig. 1.

Fig. 3.   Fig. 4.

Fig. 2.

WITNESSES.   INVENTOR.

The claim of the patent is the combination of the sight tube, the wire *SS"*, the binnacle joint and the compass as shown.

To date, no examples of this device have been identified. But this design is quite similar to one later produced by Richardson, with modification in the table and leg area, and sold by Starrett (№ 99B) and Athol Machine Company.

Ignatius S. Winter
Boston, Massachusetts

November 4, 1879
221,380

Assigned one-half to Arbee G. Hyde of Boston

IMPROVEMENT IN PENDULUM LEVELS

The Boston City Directory of 1876 showed Winter as a carpenter with a house at 87 Hudson.

Arbee G. Hyde (Arba in the directory) was said to be a conductor on the Boston & Albany Railroad (B&A R.R.) with a house at 104 Hudson. Arbee is the brother of Albe F. Hyde, who obtained patent number 1546,677

This device is a weighted wheel inclinometer. The wheel is a short cylinder where a portion of the bottom is solid but the rim is continuous. There is a finger projecting up from the base as an indicator. A spring brake is provided and actuated by pushing on a rod extending through the side of the stock and attached to wedges which, in turn, are in contact with the wedge-shaped end of the spring arm. Pushing on the rod forces the piston and spring against the rim as a brake. The cylinder is supported on an axle extending through the center of the weighted wheel. Depressions in the glass cover plates serve as bearings. The glass plates are each provided with four marks 90 degrees apart, so the device can act as a plumb or level and, with the brake, an angle once determined, can be locked in place and transferred to another surface.

No examples of this device have yet been observed.

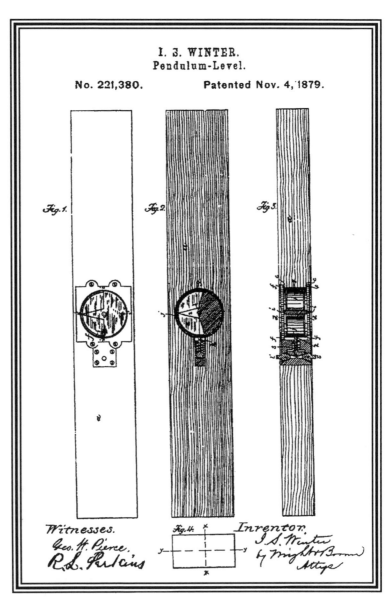

I. S. WINTER.
Pendulum-Level.

No. 221,380.   Patented Nov. 4, 1879.

Fig. 1.   Fig. 2.   Fig. 3.

Fig. 4.

Witnesses.
Geo. H. Pierce.
R. S. Perkins

Inventor.
I. S. Winter
by Wright & Brown
Attys

LEVELING-INSTRUMENT

John Warren Harmon was listed in the Boston City Directories from 1860 to 1862 as a tool maker, and he had no entry under levels at that time. He was not listed in the 1863 or 1864 Directories and that is perhaps an indication of service in the army during the Civil War. In 1865, he appears as a level maker with his business at 65 Haverhill. No occupation was stated for him in 1866 but he had a business listing of "Instrument Maker, Measuring." He appeared again in 1867 as manufacturer of spirit levels and then continuously until 1907. In 1899, his business location was given as 28 Sudbury and then later as 73 Haverhill. Harmon's spirit level business was one of the most enduring of such businesses in New England.

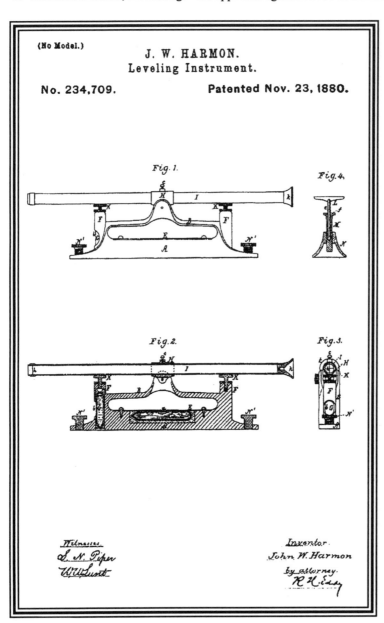

(No Model.)

J. W. HARMON.
Leveling Instrument.

No. 234,709.   Patented Nov. 23, 1880.

Fig. 1.

Fig. 4.

Fig. 2.

Fig. 3.

Witnesses.
S. N. Piper
W. W. Swett

Inventor.
John W. Harmon
by attorney.
R. H. Eddy

The patent envisions a sighting tube with provisions for the adjustment of that tube using screws located on posts marked *F*. Spirit vials arranged in the rear post and longitudinally in the base at *C* serve as plumb and level indicators. The frame has a flat base upon which is mounted the sighting tube and within which are mounted the spirit vials. Through the base are passed two adjusting screws *N'* for bringing the base into level.

It is the combination of these features that is claimed by the patent.

The instrument described by this patent is the best known of Harmon's levels and was manufactured as shown in the patent drawings. It is classified as a builders' or sighting level.

SPIRIT-LEVEL

For information on Leonard Davis, see his patent of September 17, 1867 on pages 15-16.

This is the patent for the studs entering from the rail that set the detents for the inclinometer and it is employed in many versions of Davis' levels. Only the uppermost portion of the stud shanks is threaded. The mating studs protruding from the moveable portion of the inclinometer are no longer beveled but remain cylindrical.

This patent was probably used long before the patent was granted, as it is apparently found on all Davis Level & Tool Co. inclinometers, including the mantel clock form. In one of the later versions of the mantel clock form, the studs enter from the bottom rail but in all other cases, the studs enter from the top rail.

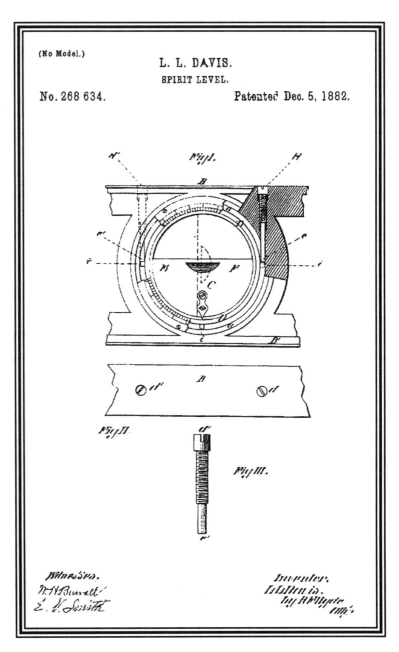

(No Model.)

L. L. DAVIS.
SPIRIT LEVEL.

No. 268 634.

Patented Dec. 5, 1882.

ARTIFICER'S LEVEL

For information on John Harmon, see his patent of November 23, 1880 on page 50.

The level is not known in the exact configuration shown in the patent drawing. The use of the table plate *O* is omitted in the versions known at this time (this is suggested in the patent language). Instead of the plate *O,* the center pivot pin shown in Harmon's patent of July 17, 1883, is incorporated. The base, with its adjustable legs and graduated scale on the side, is incorporated into the device as shown in the patent drawings. The center pin protrudes beneath the circular base and can accommodate the line from a plummet.

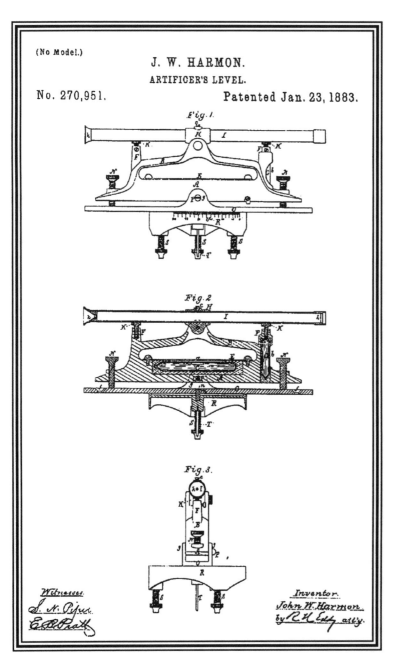

The claims of the patent include the use of the plate *O* to carry the earlier patented part of the device, and the incorporation of the circular base as shown in the drawings.

Note that in this version, there is no provision for the compass incorporated in the July 1883 patent.

An artificer is a skilled craftsman so, presumably, this level had more potential uses than Harmon's previous invention. It utilizes the same basic device patented in 1880. What has been added is a stand similar in principal to that on many surveying instruments. A large flanged top on the stand was meant to carry the sighting tube device that carries the same 1880 patent as it did as a stand-alone device.

COMBINED INSTRUMENT FOR LEVELING, SURVEYING, %C.

No information is available on Rudolph Gallis. He is never listed in the Hartford, CT City Directories.

The patent describes an instrument arranged in the shape of a standing rectangle with cross test levels on the bottom member, which is grooved to rest on a shaft. There is a sighting tube *C* along the top edge. Although the patent speaks of the sighting tube as a telescope, Gallis includes a version wherein the eye piece is a peep hole (*Fig. 4*), and the distant end is a rifle sight (*Fig. 5*). A straight edge is pivoted at the upper left corner and it may contain a spirit vial. There is a protractor scale on the two edges crossed by the free end of the straight edge. The scale may be used to determine the inclination of a piece by moving the arm until it is level and reading the angle off the intersection of the arm and frame.

Fig.6 shows a foot or stand on which the user may rest the instrument when it is not being used on shafting. The user may also use the foot to lay out lines at right angles to each other.

The instrument envisioned by this patent is for use primarily as a shafting level. To date, no examples of an instrument utilizing this patent are known.

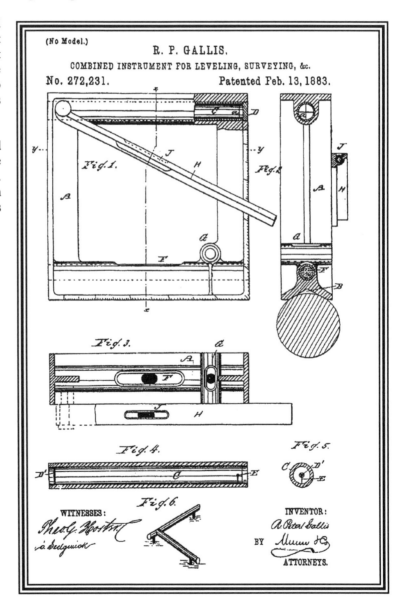

-53-

John W. Harmon
Boston, Massachusetts

July 17, 1883
281,267

ARTIFICER'S LEVELING INSTRUMENT

For information on John Harmon, see his patent of November 23, 1880 on page 50.

The patent calls for the use of a tripod with a special top plate to mate to the three legs of the base. The patent also describes an index pointer *i* fastened to the top of the sighting tube in such a way as to assist in determining the compass reading for the line of sight.

This level builds on patent № 270,951 by adding a compass box on top. The stand changes some to accommodate the split legs of the compass box, but is generally similar to that used in the January 1883 patent device. This form of Harmon's patent is known.

PATENT DRAWINGS ON NEXT PAGE.

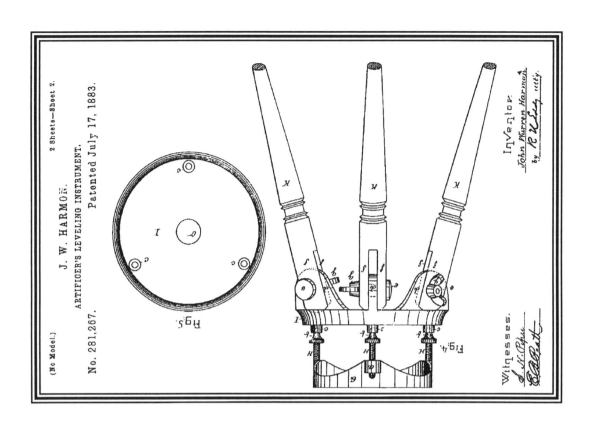

(No Model.)

2 Sheets—Sheet 2.

J. W. HARMON.

ARTIFICER'S LEVELING INSTRUMENT.

No. 281,267.

Patented July 17, 1883.

Fig.5.

Fig.4.

Inventor:
John Warren Harmon
by R. Wiley atty.

Witnesses.

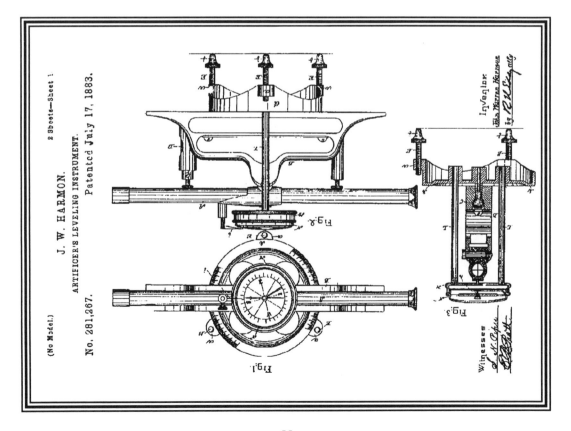

(No Model.)

2 Sheets—Sheet 1

J. W. HARMON.

ARTIFICER'S LEVELING INSTRUMENT.

No. 281,267.

Patented July 17, 1883.

Fig.2.

Fig.3.

Fig.1.

Inventor:
John Warren Harmon
by R. Wiley atty.

Witnesses

BEVELING INSTRUMENT

For information on Laroy Starrett, see his patent of May 6, 1879 on page 47.

This patent describes a device consisting of a grooved rule in combination with a stock having a straight edge, A, a revolvable divided head that would accept the rule, and a clamping screw to hold it firm. One of the claims was that the stock should have a spirit level.

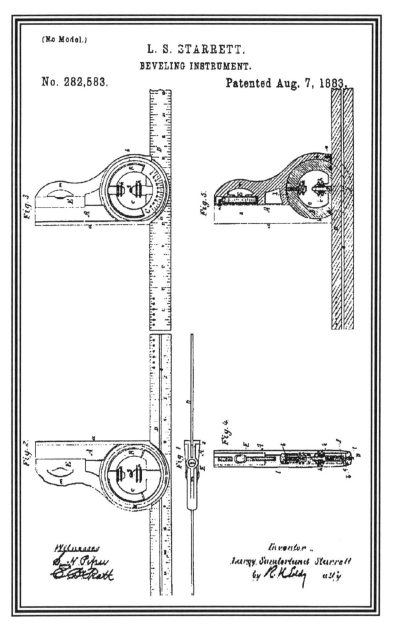

(No Model.)

L. S. STARRETT.
BEVELING INSTRUMENT.

No. 282,583.

Patented Aug. 7, 1883.

This device is very well known. The beveling head, in combination with a metal rule, formed a high quality inclinometer with a spirit level in parallel with the base of the head. The rule fits into a rotating protractor in the head that contains a graduated scale. This was sold as Inclinometer Nº. 10.

Nelson H. Bearse
Osterville, Massachusetts

SPIRIT-LEVEL

No information is currently available about Nelson Bearse.[13]

This patent incorporates a pair of mirrors installed above the spirit vial in the level fixture — one on each side of the opening — that enables the position of the bubble to be read from below on either side of the stock. The plumb porthole is to have a convex mirror located opposite the spirit vial, again to allow the position of the bubble to be read from below on either side of the stock. One of the aims of the patent is to facilitate reading the bubble from anywhere without moving the mirror.

Wilkinson (in Boston) advertised Chapin levels for sale outfitted with this invention. An example was not available for study.

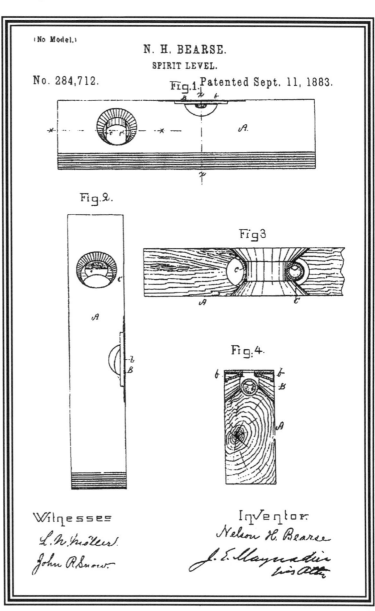

13   Osterville library closed for remodeling and unavailable for research during the study period for this book.

-57-

LEVEL

Born in 1839, Van Alstyne was a prominent citizen of the Sharon area, and he was in the Connecticut legislature during 1895-96. In 1910, he wrote book about his service during the Civil War. His principal business was said to be carpentry and building; he was also superintendent of the Sharon Valley Malleable & Gray Iron Co.

Van Alstyne's patent envisions an annular tube, half filled with liquid, set in a circular opening in a wooden stock. Such a level requires that the annular tube be exactly half full. There might be graduations around the outer edge of the tube in the wood. When the tube contains graduations, it functions as an inclinometer.

To date, no examples of an inclinometer or level utilizing the features of this patent are known.

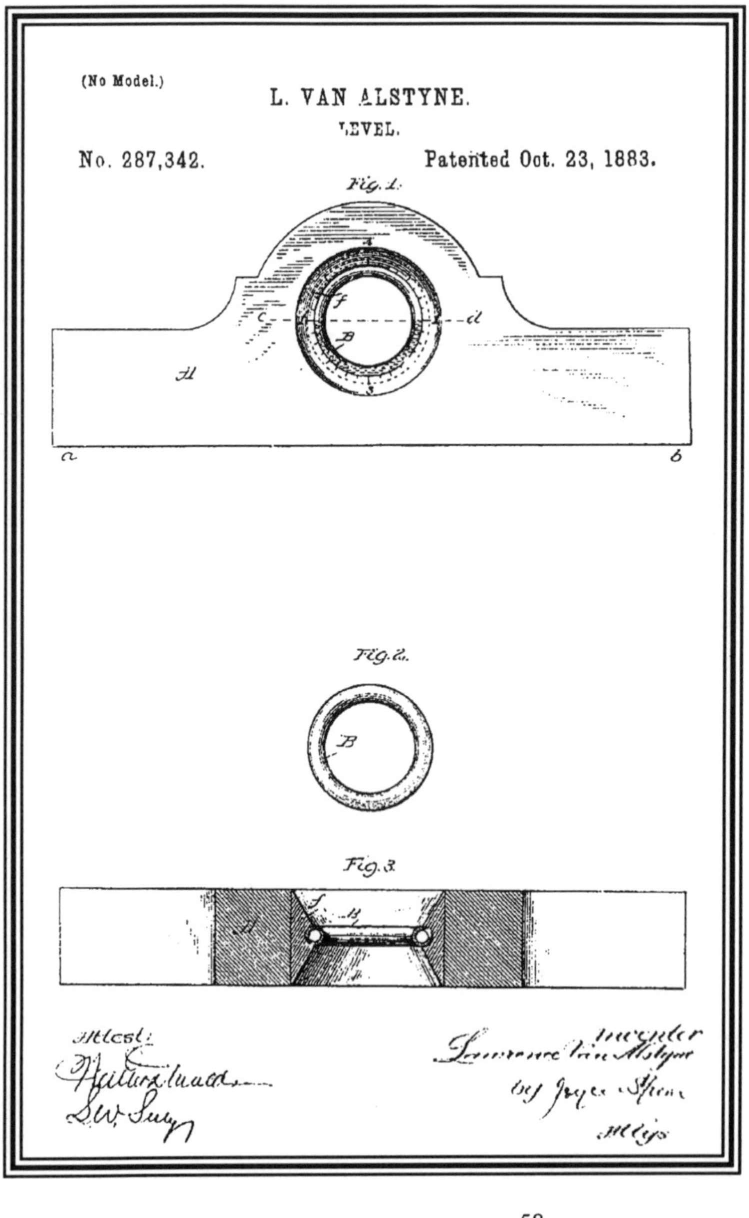

(No Model.)

L. VAN ALSTYNE.
LEVEL.

No. 287,342.   Patented Oct. 23, 1883.

Fig. 1.

Fig. 2.

Fig. 3.

SPIRIT-LEVEL AND PLUMB

For information on Leonard Davis, see his patent of September 17, 1867 on pages 15-16.

The patent calls for studs passing through the bottom rail and then through springs and into threaded holes in the plugs at the end of the level cartridges. The adjustment can be made by turning the studs in opposite directions.

The plumb tube utilizes an eccentric device similar to that in the Davis 1871 patent for the adjustment of the plumb fixture in wooden carpenters' levels. Turning the plug at the base of each tube, by using a slot cut into the bottom side of the plug, causes the base of the tube to move sufficiently to adjust the plumb calibration.

This is the patent used on Davis metal carpenters' levels to adjust both the level and the plumb vials. The drawing shows the 6" version of the so-called spindle form level that was the first design for the metal carpenters' levels.

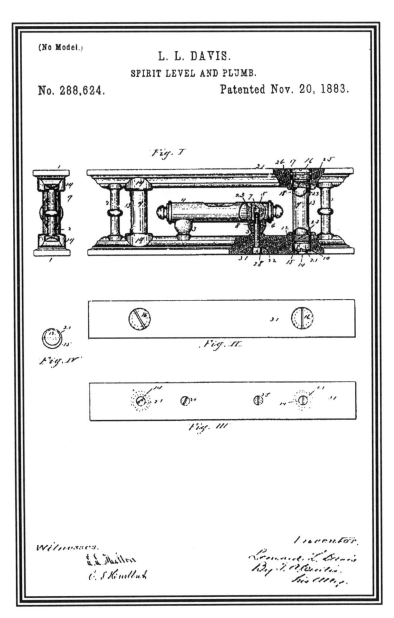

(No Model.)

L. L. DAVIS.
SPIRIT LEVEL AND PLUMB.

No. 288,624.   Patented Nov. 20, 1883.

Stephen H. Bellows
Athol, Massachusetts

COMBINED BEVEL AND T-SQUARE

Steven Bellows was associated with Standard Tool Co for many years. After the purchase of the Standard Tool Co. by Starrett, Bellows retired to Shrewsbury. He was the patentee on many of the patents obtained by Standard Tool as it struggled to keep up with Starrett.

This was Standard Tool Co's. answer to the Starrett's patent No. 282,583. It can be used independently as an inclinometer by removing the rule and inserting the piece carrying the spirit vial (*Fig.2*) into the revolving head on the stock. It was sold as "Bevel Protractor (with Level Attachment)".

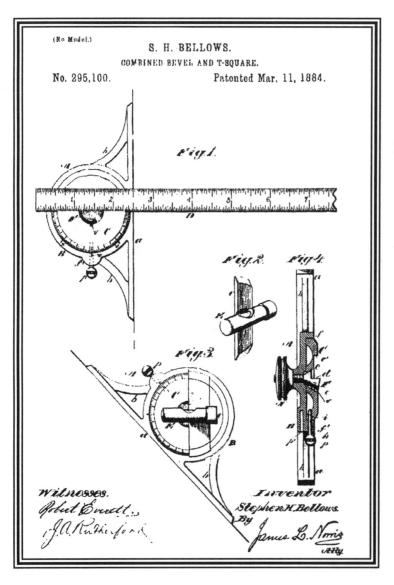

This piece was eventually manufactured and often the level piece is found separated from the rest of the tool. Starrett sued for patent infringement and won, so this tool was not produced until 1897 when the Starrett patent had expired.

Charles J. Parkhurst & Albert W. Parkhurst
North Adams, Massachusetts

April 8, 1884
296,608

Assigned 100% to Charles J. Parkhurst of North Adams, MA

PENDULUM-LEVEL, PLUMB AND INCLINOMETER

In various North Adams City Directories, Charles Parkhurst was recorded as a law student in 1874-1875. By 1884, Charles was an attorney in the firm of Parkhurst and Couch, while Albert W. Parkhurst was a mason at Natural Bridge.[14] From 1879 through 1888, Charles is listed as an attorney, but Albert (probably the father of Charles) was not listed other than in 1884.

This is a very complex tool. A pendulum hangs in a U-shaped channel that is lined with elastic materials or flexible plates. A wedge, shown as *K* is controlled by the lever in the side of the stock. When the wedge is pushed to the right, the channels are forced apart and the pendulum can swing. When the wedge is withdrawn, the channel walls grip the pendulum holding it in position. The pendulum is attached to a bevel-gear that rotates in a vertical plane. The beveled teeth engage a second bevel-gear rotating in a horizontal plane. The second gear is attached to the shaft of a needle rotating over a dial on the top surface of the level and graduated from 0 - 90° degrees. The gear arrangement thus provides a multiplying factor of four.

The bottom of the stock may be a metal plate, in which case there would be a metal band on three sides.

To date, no examples of this device have been located.

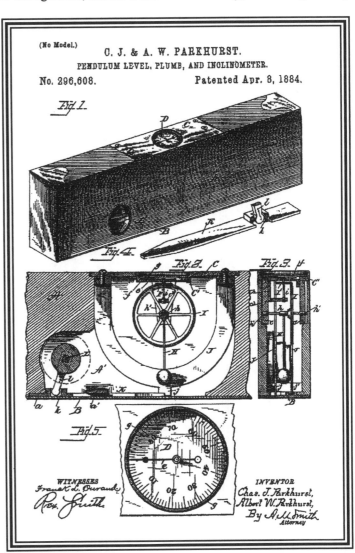

---

14   According to the Gazetteer of Berkshire County, Mass, Part Second, Business Directory of Berkshire County Mass. 1884-85 compiled and published by Hamilton Child.

SPIRIT-LEVEL

For information on Leonard Davis, see his patent of September 17, 1867 on pages 15-16.

This patent envisions putting threaded studs through the top and the bottom rail, and into the plugs on the ends of the level case. The claim is for rigidity and ease of vial replacement because the case need only be freed on one end and then could be swung out for easy access.

No example of the level envisioned by this patent has been located. However, the concept is similar to a patent granted to, and used by, Goodell for Millers Falls (№ 310,046).

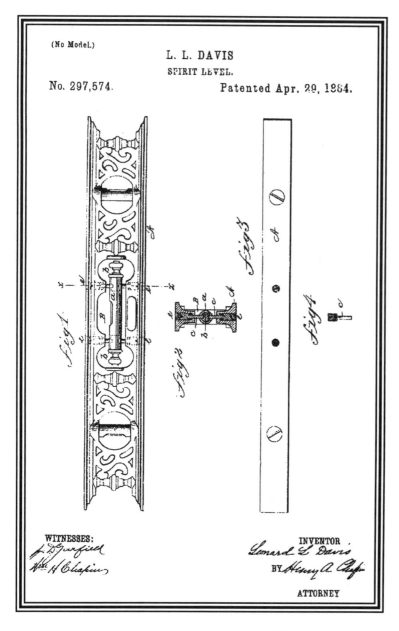

Albert Root
Hamden, Connecticut

April 29, 1884
297,719

COMBINED LEVEL AND BORING-GUIDE

Albert Root was a carpenter and joiner. After obtaining this patent, he left Hamden for Pennsylvania and remained there for the remainder of his life.

Root's patent is for a rectangular block that contains a small inclinometer on each of two adjacent sides. The tool is meant to attach to a bit shank using a set screw. The inclinometers are weighted pointers and around each is a graduated dial (indicator ring). The graduated rings can be preset to any angle for boring slanted holes.

To date, there are no known examples of this tool.

PATENT DRAWINGS ON NEXT PAGE.

A. ROOT.

COMBINED LEVEL AND BORING GUIDE.

No. 297,719. Patented Apr. 29, 1884.

2 Sheets—Sheet 1.

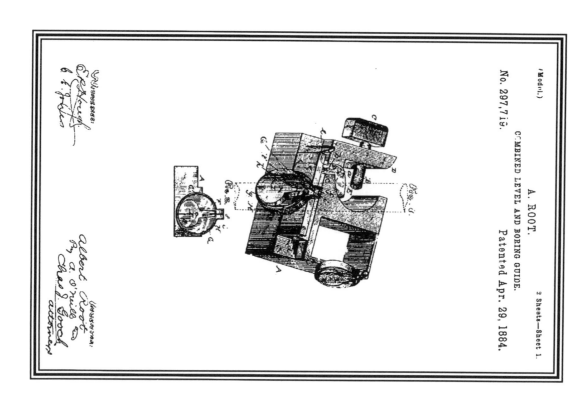

A. ROOT.

COMBINED LEVEL AND BORING GUIDE.

No. 297,719. Patented Apr. 29, 1884.

2 Sheets—Sheet 2.

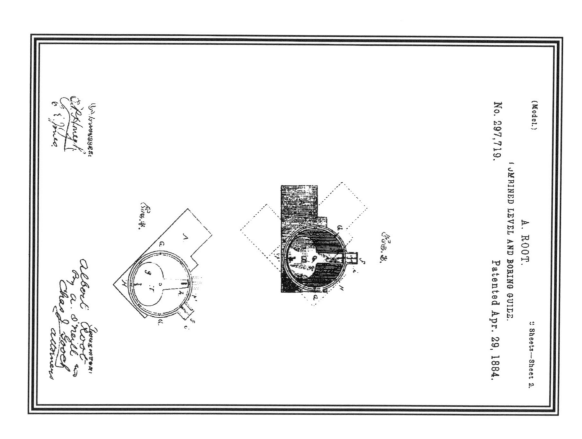

F. W. Ritchie
Vanceborough, Maine

August 19, 1884
303,666

Assigned two-thirds to John Ritchie and Michael L. Ross of Vanceborough, ME

COMBINATION-TOOL

Frederick W. Ritchie was a 30 year old house carpenter at the time that he obtained this patent. He was born in New Brunswick, Canada.

There is no census listing for Michael L. Ross in 1880 or 1900, and no other records from Vanceborough could be located.

The patent describes a try square containing a bevel blade in the stock opposite the fixed blade. Spirit vials are set into the stock in both plumb and level positions.

The Ritchie Patent bevel was manufactured and this tool has been made from various combinations of materials. It is a very collectible tool.

The tool, as manufactured, differed from the patent drawing in that: (1) in the manufactured tool a protective cover is provided for the plumb vial; and, (2) in the patent drawing there is a slot *s* at the point that the bevel is held, that will allow the bevel tongue to be moved toward the inner edge of the tool when desired. This slot is not found on the available versions of the tool.

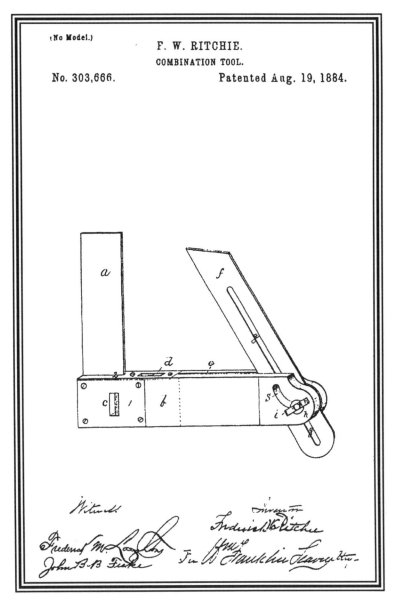

Henry Murdock Rich
Athol, Massachusetts

October 14, 1884
306,429

Assigned to C.F. Richardson of Athol, MA

INCLINOMETER

H. M. Rich was a machinist who apparently worked for Richardson. This is interesting primarily because L. D. Rich (apparently the father of Henry) was a prominent figure in Athol Machine Co. Richardson will be covered elsewhere under his own patents.

As envisioned by the patent, a coarse gear is fixed to a knurled screw. When the screw is turned the teeth mesh with teeth on the base of a semicircular segment containing a spirit vial. The frame and moving piece are very similar to the same parts of Davis's inclinometer. The teeth are found on a bottom quadrant of the moveable semicircular portion. Turning the screw causes the level to rotate through 90°, just as is done manually in a Davis inclinometer.

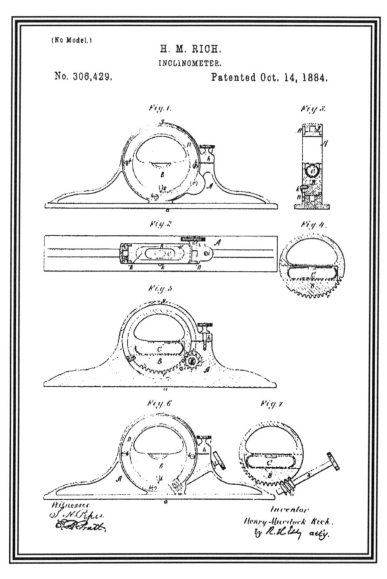

Known examples of a level according to this patent have the inclinometer set in a 12" cast iron level stock much as a Davis. None of the mantel clock form, as envisioned by the patent drawing, are known. The known examples have a round thumb screw on both side of the stock, with one screw moving inclinometer part in one direction, and the other screw moving the inclinometer in the opposite direction.

Albert D. Goodell
Millers Falls, Massachusetts

December 30, 1884
310,046

Assigned to the Millers Falls Co. of Millers Falls, MA

SPIRIT-LEVEL

Albert D Goodell was born in Whitingham, VT, and he learned the trade of carpenter.[15] In 1870 he went to Millers Falls, MA, to the Millers Falls Company, becoming an inspector, superintendent and master mechanic. He had several patents to his credit during his time there. In 1888, he went to Shelburne, MA, where, with his brother Henry E., he formed Goodell Brothers. This association lasted for four years, after which he sold his interest to Henry. (Henry soon moved that business to Greenfield where, with another brother Dexter, he formed, what became, Goodell Pratt.) Albert moved to Worcester, where he formed the Goodell Tool Company with his son, Frederick. They stayed in Worcester for about a year and then returned with the business to Shelburne. This excellent business was taken over by Goodell-Pratt in 1918. (Frederick Goodell established a small factory and manufactured small tools until his death in 1929.)

The patent calls for a central level cartridge to be held by two opposed, sliding studs on each end. The studs are held in place by short screws through the frame. One stud of each pair bears against the outside of the level cartridge while the other bears against the end plug inside the cartridge. Adjustment is accomplished with the screws in the frame. This patent requires two cone shaped studs per plumb for adjustment as shown in the patent drawing.

This patent was used as the basis for the early Millers Falls adjustable metal levels. It is similar to Patent Nº 297,574 issued to Davis earlier in the same year.

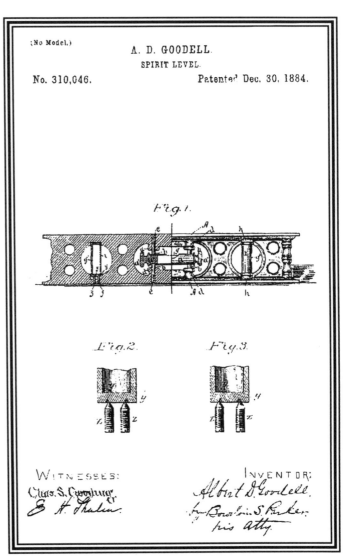

(No Model.)

A. D. GOODELL.
SPIRIT LEVEL.

No. 310,046.   Patented Dec. 30, 1884.

Fig. 1.

Fig. 2.   Fig. 3.

WITNESSES: INVENTOR:
Chas. S. Cushing.   Albert D Goodell.
E. H. Thurlow.   by Bowdoin S. Parker,
                             his atty.

15   History and Tradition of Shelburne, Massachusetts, History and Tradition of Shelburne Committee, 1958

COMBINED SQUARE, BEVEL, AND LEVEL

Bisbee Merrill was born in Brewer, Maine in February 1850. In the 1880 census, Merrill was listed as a clerk in a drug store. In 1886, he opened his own drug store. He was mayor of Brewer in 1894 and 1895 and was elected a Country Commissioner in 1900. In the 1900 census he was listed as a drug merchant.

The patent for this device describes a try-square with its tongue being an angle iron-like piece. There is a moveable blade (bevel) in the end opposite the fixed blade. The moveable blade contains a notch in the end to engage the side a of the angleiron and, when fully inserted, is locked into the stock. The level vial is in a housing or cartridge in the stock and is connected to a rod that extends to the end of the stock. At the external end of the rod are wings that can be turned to rotate the vial cartridge. This has the same effect as putting a cover over the spirit vial.

No example of this tool is known at this time. This device is similar to that described by Ritchie's Patent and may have been inspired by that patent.

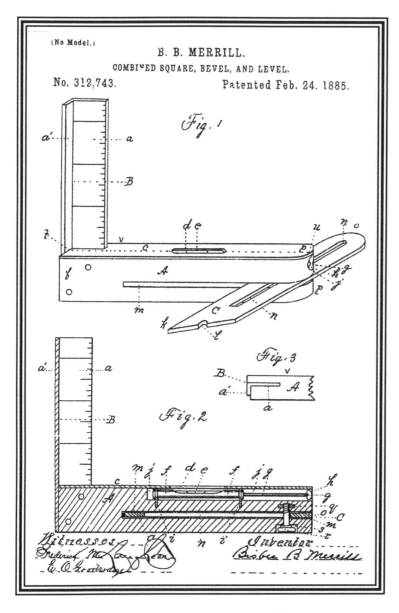

(No Model.)

B. B. MERRILL.
COMBINED SQUARE, BEVEL, AND LEVEL.

No. 312,743.   Patented Feb. 24. 1885.

Fig. 1

Fig. 3

Fig. 2

Witnesses   Inventor
                             Bisbee B Merrill

COMBINED LEVEL SQUARE AND BEVEL

Matthew Euington was an Englishman, and an expert patternmaker and whittler.[16] He and his entire shop were brought to Upton by William Knowlton. (Knowlton operated a straw hat manufacturing business in Upton.) Euington was born in 1834 in England and died in 1902 in Upton.

Part of the device, described by this patent, is an ordinary try-square with a spirit level vial in the handle. To this is added an adjustable segment gage with a provision made to fasten it, with a thumb screw, in either of two positions on the blade. (Only one segment gage was envisioned, and the patent drawing shows that one gage in all of the possible positions. The third position (along the blade) was provided for storage of the bevel segment when not in use. Grooves in the blade are provided, in each position, to stabilize the gage piece.

The gage and its associated groove have a radius of curvature of one foot and each groove is located exactly one foot g from the opposite end of the blade. Thus by moving the gage piece in or out, any inclination to a foot may be determined.

No example of this device has yet been observed.

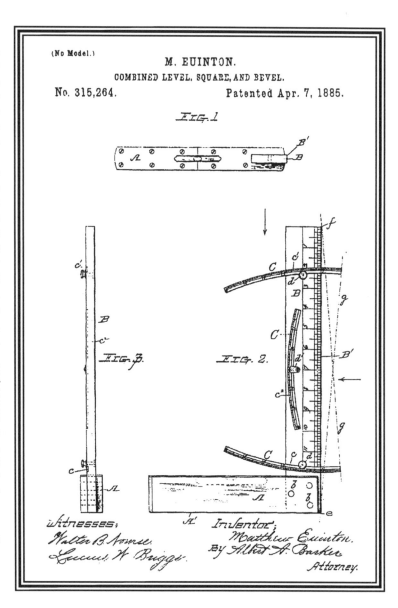

---

[16] *Upton's Heritage, the History of a Massachusetts Town* by Donald Blake Johnson, 1984, Phoenix Publishing Co., Canaan, NH

James Walsh, Thomas E. Murphy and Everett A Clark   May 5, 1885
North Adams, Massachusetts   317,250

Clark assigns to Walsh and Murphy

LEVEL

E. A. Clark was listed as a painter in the 1879 North Adams City Directory and as a watchmaker and jeweler from 1883 through 1886. He was not listed in the city directories in 1887 or 1888. The Gazetteer of Berkshire County[17] listed Clark as a watchmaker and engraver with a business on Main Street and a house at 16 Pleasant St.

The Gazetteer also showed James Walsh as a contractor and builder with a house at 5 Union St. Walsh was first listed in the 1883 North Adams City Directory.

No plausible listing was found for Murphy until the 1887 North Adams City Directory, when a Thomas Murphy was listed as a printer at the Hoosic Valley News and in the 1888 directory as a printer at Adams Transcript. However, a Thomas E. Murphy was listed in the North Adams City Directory as Asst. Pastor of St. Francis Roman Catholic Church.

The patent envisions an inclinometer housed in a standard wooden stock into which a handle has been inserted near the top center. A chamber in the center of the side of the stock houses the working parts of the inclinometer and it is provided with a hinged cover. A closed cup, containing mercury, is provided behind the dial. A toothed float on an axle through the center of the cup transmits the inclination of the surface of the mercury to a pointer in front of the dial. The float is toothed in order to provide room for expansion from temperature fluctuations. There is a geared wheel, on the same shaft as the pointer and the float, and it meshes with a much smaller gear on a second shaft. The second shaft also carries a pointer that is to be read against a second scale on the face. The ratios of the diameters are such that the small gear makes 24 revolutions for every one revolution of the large gear. In order to avoid backlash, a second wheel is loosely mounted on the main shaft and also in contact with the small gear. This second wheel carries a spring, one end of which is attached to the wheel and the other end of which is attached to the shaft as shown in *Fig.6*. The two large wheels are in contact and thus backlash is prevented.

Fifteen degrees on the big dial translates to one full revolution on the small dial and the small dial is accurately graduated to tenths of degrees. Thus a precise angle measurement is said to be possible.

No examples of this device are known at this time. Its proposed operation and design certainly speak for the involvement of a watchmaker as one of the inventors.

PATENT DRAWINGS ON NEXT PAGE.

---

17   *The Gazetteer of Berkshire County, Mass, Part second, Business Directory of Berkshire County, Mass 1884-85.* (Compiled and published by Hamilton Child).

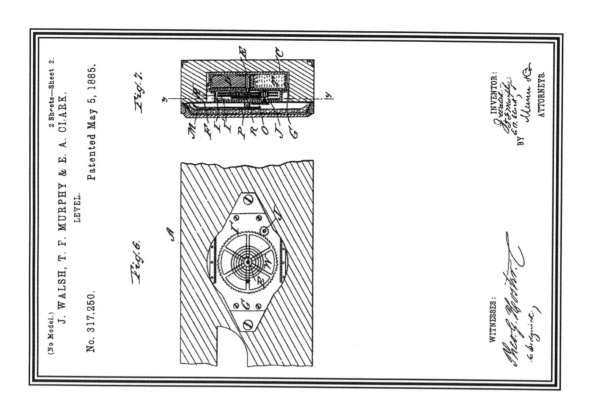

Fig. 6.　　　　　Fig. 7.

WITNESSES:

INVENTOR:

BY
ATTORNEYS.

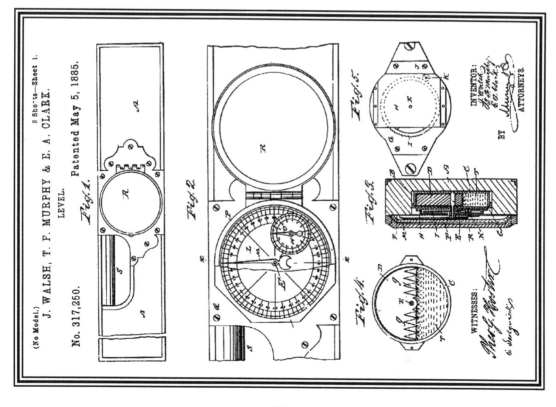

Fig. 1.

Fig. 2.

Fig. 3.

Fig. 4.

Fig. 5.

WITNESSES:

INVENTOR:

BY
ATTORNEYS.

COMBINATION SQUARE AND GAGE

No information is available regarding William Ford[18].

The patent calls this a combination square & gauge. It is, in fact, a machinists' combination level, bevel, try square, T-square, surface scratch gage, depth gage, and thread & drill gage. It is obvious from the patent drawing that the spirit vial in the handle of this tool in its bevel/try square format would function as a level.

There are no known examples of this tool. However, it seems that the tool, as designed, has too many parts to have survived intact even if it was ever produced.

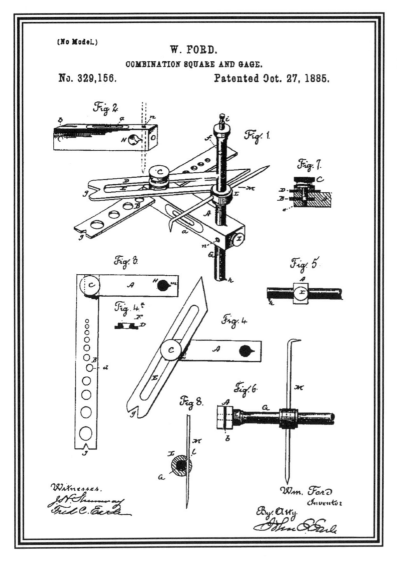

(No Model.)

W. FORD.

COMBINATION SQUARE AND GAGE.

No. 329,156. Patented Oct. 27, 1885.

Fig. 2. Fig. 1. Fig. 7. Fig. 3. Fig. 5. Fig. 4.ᵃ Fig. 4. Fig. 8. Fig. 6.

Witnesses.

Wm. Ford
Inventor

By. Atty.

---

18   The library serving Birmingham (Derby) during this study was undergoing substantial renovation and the historical documents were unavailable to the author.

COMBINED RULE, CALIPERS, AND LEVEL

Carsten Janssen first appeared in the 1884-85 Torrington City Directory as an architect (the only one in town) with offices at 211 E. Main. From 1885 to 1890, he was employed by Hotchkiss Bros. & Co. (Carpenters, Builders & Supplies) and subsequently he formed Beckley & Janssen (Carpenters & Builders). By 1899, he was listed only as an architect and in the 1903-04 Torrington City Directory he was recorded as having moved to New York.

The Janssen patent is for a combined rule, parallel rule, level and calipers and that can be ascertained from the patent drawing. The patent, however, maintains that the tool is "..an ordinary jointed rule, a parallel ruler, a compass, a protractor, a square, a scale and quadrant, a plumb, and a level all combined in a simple, compact and convenient manner ..."

The spirit vial can be observed, in *Fig.4*, on the inside of the right hand leg.

To date, there are no known examples of this tool.

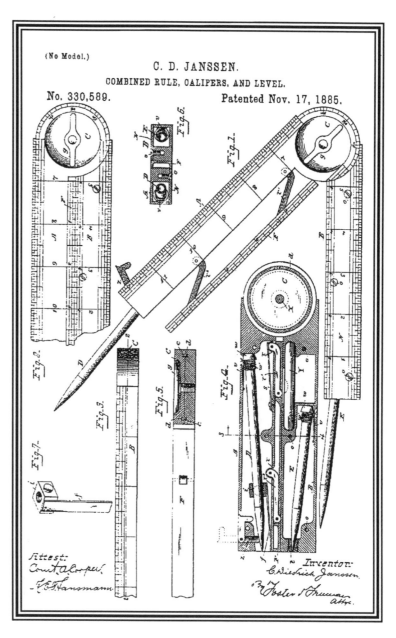

L. Arthur Sanford
Bristol, Connecticut

March 9, 1886
337,621

Assigned two-thirds to Leverett A. Sanford and Rufus A. Sanford of Bristol, CT

SPIRIT-LEVEL

L. Arthur Sanford or Leverett A. Jr. and the other Sanfords worked at Dunbar Bros., a maker of clock springs. L. Arthur was listed as employee in 1886 and 1888; a foreman in 1895; and a master mechanic in 1908. Leverett A. Sanford (apparently the father) was shown at Dunbar Bros. as a foreman in 1886, a toolmaker during 1895 through 1897, and as a machinist in 1908.

The tool described by the patent is an inclinometer, where a chamber is partly filled with a fluid (mercury) and the inclinometer needle and a float are on the same arbor. The mercury and the float are in a sealed chamber behind the dial while the needle is, of course, in front of the dial. Movement of the mercury causes the float to move and, in turn, move the indicating needle.

It was thought that mercury would provide a faster reaction time (it would) and would be less affected by atmospheric temperature and pressure (also true).

A cover is also to be provided.

To date, there are no known examples of this inclinomter.

COMBINATION SQUARE, LEVEL, AND BEVEL

According to the census records of 1900 and vital statistics records, John Finley immigrated from St. Andrew, New Brunswick, Canada in 1863, but was never naturalized. He worked as a house carpenter. He was born in June 1847 and died April 24, 1921.

This device comprises a stock containing level and plumb fixtures, with a pivoting tongue at one end that may serve as a rule, square, or bevel. For use as a square, the tongue folds into a mortise which positions it at 90° to the stock. A sliding bolt F inside of the stock is moved by pushing at k and can be used to lock the tongue in position when used as a square.

No example of this device is known to date.

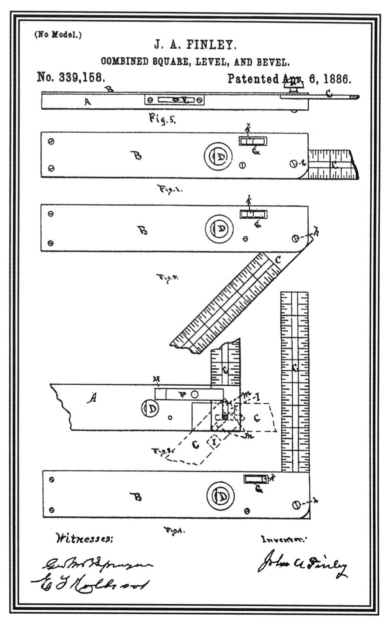

TRY AND BEVEL SQUARE

No information about Ambrose Vose was located.

This device was envisioned as being composed of several pieces. The basic stock contains a spirit vial and a graduated disk capable of locking-in a machinist's level. The second piece is a sliding bar *Y* containing a scratch awl *E*. A third piece is a sliding bar *T* that fits onto a rule and can be locked thereon. Claim is made that the device can be used as a center-finder and a compass when the pieces are properly arranged.

To date, no example of this device has been located.

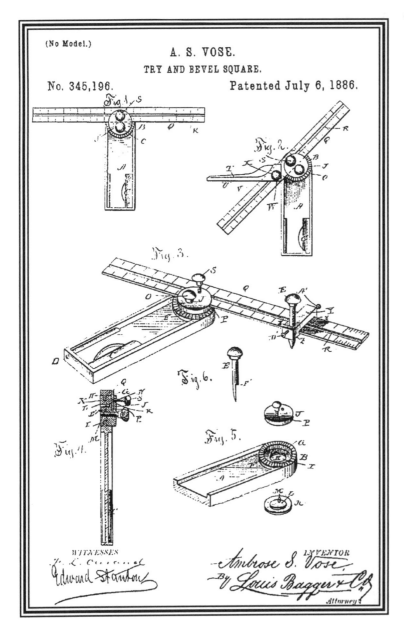

Assigned to the Stanley Rule and Level Company of New Britain, CT

LEVEL ATTACHMENT FOR BORING TOOLS

For information on Justus A. Traut, see his patent of October 6, 1868 on page 28.

The patent describes a circular frame carrying a spirit vial. The shaft of a tool can rest in grooves on the rear side of the frame. A screw clamp, behind the frame, holds the device onto the shaft of a tool such as a bit. The blade of a square can be held to the back of the frame by the clamp. All that is new about this invention is the provision for more than one "socket" to hold the bit. The sockets are at useful fixed angles, i.e., level, plumb and 45°.

This patent resulted in the Stanley № 44 bit and square level. It has been manufactured in exactly the form shown in the patent drawing. The idea of a bit with a square level was not new, nor was the idea for a demountable version of such a level.

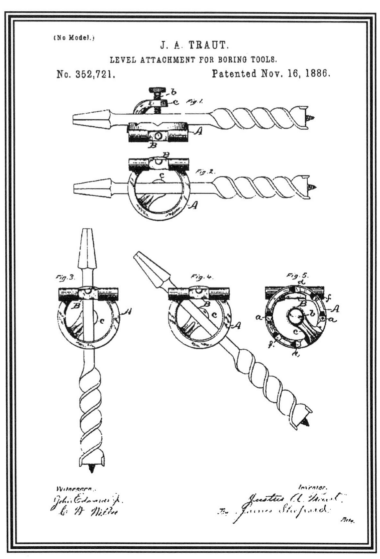

(No Model.)

J. A. TRAUT.

LEVEL ATTACHMENT FOR BORING TOOLS.

No. 352,721.   Patented Nov. 16, 1886.

Edward E. Webb
Fitchburg, Massachusetts

December 7, 1886
354,076

SPIRIT-LEVEL

Edward Webb first appeared in the Fitchburg, MA, City Directories as a partner in Webb and Stone (Fitchburg Level Co.). His partner, Hosea Stone, had been employed by the Heywood Chair Mfg. Co in 1885 and 1886. The directories indicate that Edward Webb went to Los Angeles, California, in 1887 or 1888, and it appears that he took no further interest in the manufacture of his patent.

In this patent all three spirit vials are placed in the same central compartment and cemented in place. The level shown in *Figs. 1 - 5* was not proposed to have side plates. The space between the plumb vials and under the level vial would be open and could act as a handle of sorts. One of the statements in the patent is that this form of the level did not require any screws, holes or special holders. Another is that any form of ornamentation between the rails is OK. Apparently Webb thought better of this idea and proposed the modification shown in *Fig. 6* incorporating both side plates and a specific form of ornamentation.

This patent is the basis for the Fitchburg Level Co. cast iron levels. This level was manufactured for several years by a succession of at least four companies, beginning with the Fitchburg Level Company and ending with Empire Level Company. All examples are faithful to *Fig. 6* of the patent drawings.

PATENT DRAWINGS ON NEXT PAGE.

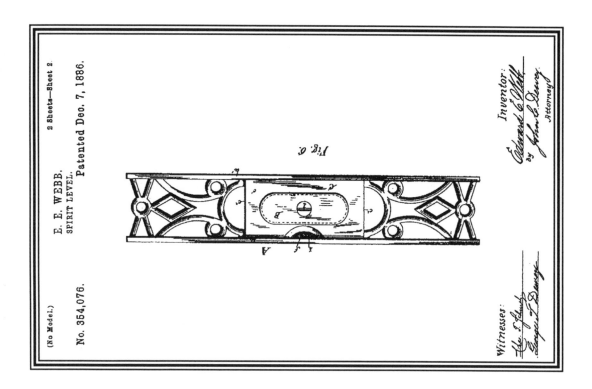

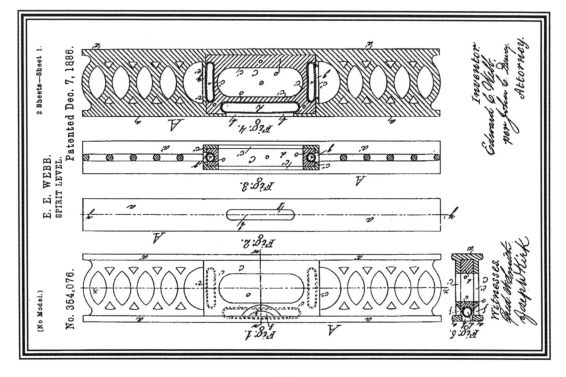

William Wood Brown
Manchester, New Hampshire

December 21, 1886
354,592

INSTRUMENT FOR PLUMBING THE CORNERS OF BUILDINGS

William Wood Brown worked as a laborer in 1886.

This device comprises a base plate with two threaded clamps that hook into the joints in a masonry wall, one on each of two adjacent sides. On this base sits another table fitted with cross test type level vials in a cross test arrangement, and three or four adjustable legs *f*. The legs are to be threaded, and are used to level the table. A graduated standard (rule or blade) is set vertically into the top table and when the table is level, the standard should be plumb. This is meant to accurately define the plumb corner of a masonry wall.

To date, no example of this device has been identified.

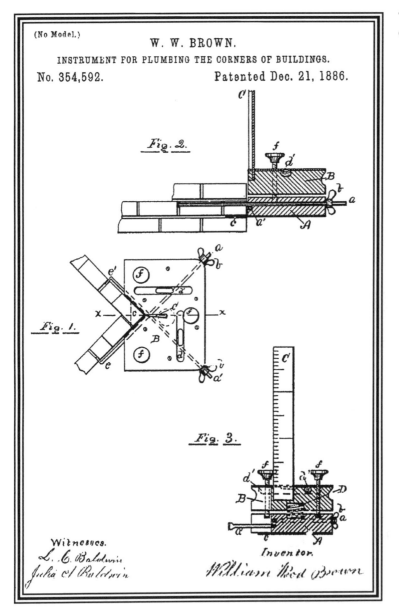

PLUMB-LEVEL

Edward Duffy first appeared in the Providence City Directory in 1888. He was listed first as being a machinist with a business address at 12 Johnson. In 1889-90, he had a house at 31 Pawtucket (and one at a different location every year to 1894). In 1895 and 96, Duffy Bros. (Edward C. and Thomas H.) were listed as electricians. In 1896, they (Edward C and Francis J) were also listed as being in the hardware business at 40 East Ave. In 1897, that business listing became Duffy Bros and Co. (Edward C., Francis, and William K. Toole) This hardware business continued beyond 1915 and at some point Francis Duffy left and Thomas Duffy returned.

The general style of the inclinometer proposed in this patent is of a weighted needle type. The index part of the needle extends in front of the dial, while the weight is behind the dial but on the same shaft. The patent calls for a post *11* at the front of the device; upon it one end of the axle will bear. The other end of the axle bears on the end of a thumb screw protruding through the frame from the rear. The thumb screw may be tightened to act as a brake or a lock.

Levels made generally according to the specifications of this patent are known.

The actual level omits the front post and apparently bears directly upon the glass. In the rear is found a small slotted screw which most probably serves the same function as the thumb screw, but obviously is not intended for frequent adjustment.

The faces of the known levels appear to be silvered and are highly reflective. The patent claims that by knowing the length of the pointer and using a logarithmic table of tangents, the relative inclinations can be determined, i.e, in terms of rise to run. The actual level dial is also calibrated in radians, but the zero points of the scales do not mesh.

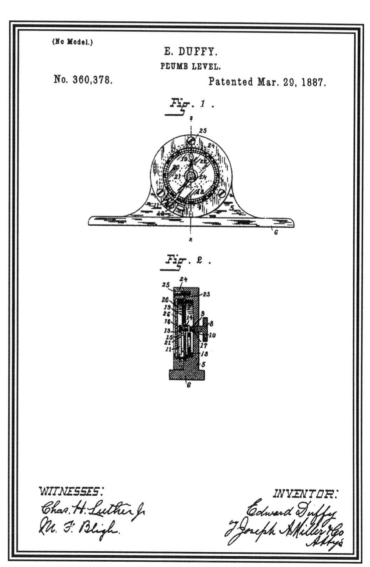

SURVEYOR'S TRANSIT

Charles Frederick Richardson was born to Nathaniel and Emerline Richardson on September 28, 1839. Nathaniel operated a general machine shop from 1835 until 1883; however, in later years the shop was run by George and Charles Richardson. It was here, in 1878, that Starrett rented space to make his combination square. In 1880, under threat of legal action from Athol Machine, Richardson severed ties with Starrett and sold him a portion of the shop to make his square. Upon his father's death, he bought out his brother. He died April 19, 1917.

The patent in question is interesting for the fact that (1) the cast iron level is described as a stand with space left in the top center for a level vial; (2) the reference to the top tube as a telescope, and (3) the three claims of the patent all concern the tripod with its graduated table. The "stand" is, in fact, a stock Richardson cast iron level with holes drilled into it to accommodate supports for the tube on top and to fit to the table on the bottom. The tube, in all known examples, is a sighting tube with pin hole and cross hairs and no optics. The table, as known, has the graduations and the extendable legs.

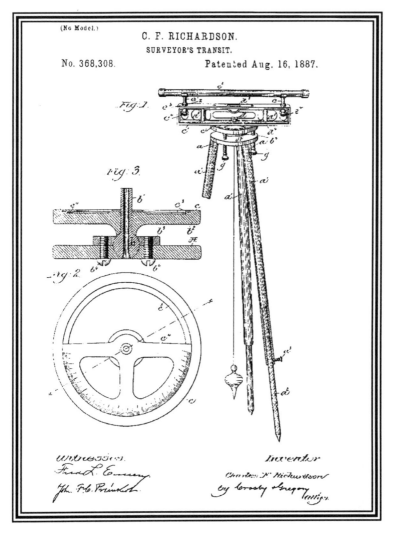

This tool was manufactured by Richardson, in a form that was faithful to that described in the patent and shown in the drawing.

Charles Tiller
Milford, Massachusetts

August 16, 1887
368,434

BEVEL

Tiller was a British citizen at the time of the patent. In 1886-87, he was a stonecutter for Norcross Bros. which ran a granite quarry near Rocky Woods, Massachusetts. Norcross Bros. was a builder/contractor from Worcester involved in erecting major buildings of brownstone or granite. After 1890, Tiller was no longer listed in the Milford City Directories.

The device is a double-tongued bevel tool with provisions for either one at each end of the stock, or both at the same end. The tongues are secured by thumb screws and the ends of the stock, containing the thumb screws, are graduated in 15 degree increments. The center of the stock is identified and provision is made to fold the arms into an isosceles triangle to hold a plumb bob, such that the device can be used as a plumb level as shown in *Fig.4*. Many other uses of the tool as a bevel, square or rule are indicated.

Examples of this device have not yet been found for examination.

PATENT DRAWINGS ON NEXT PAGE.

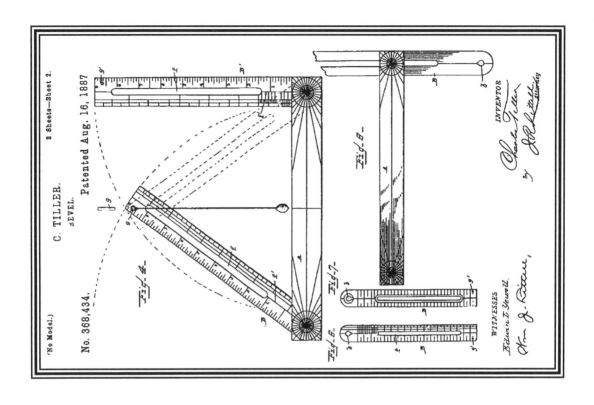

C. TILLER.
BEVEL.

No. 368,434.   Patented Aug. 16, 1887.

Fig. 4.

Fig. 6.

Fig. 7.

Fig. 8.

WITNESSES
Edwin I. Yewell.
Wm. J. Rowe,

INVENTOR
Charles Tiller
by J. R. Littell
Attorney

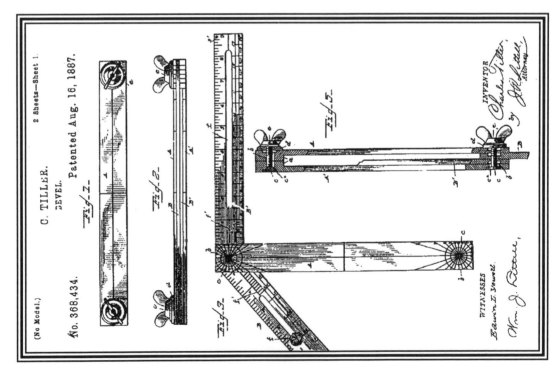

C. TILLER.
BEVEL.

No. 368,434.   Patented Aug. 16, 1887.

Fig. 1.

Fig. 2.

Fig. 3.

Fig. 5.

WITNESSES
Edwin I. Yewell.
Wm. J. Rowe,

INVENTOR
Charles Tiller
by J. R. Littell
Attorney

-84-

SPIRIT-LEVEL

Philo Fox was the son of George Fox, a carpenter. Philo also worked as a carpenter at W. A. Wilson, beginning in 1881. In 1887, he was shown as a partner in Fox & Callaghan (whose business was as carpenters and builders). In 1891, he was shown as a builder with A. W. Burritt and, eventually, as Vice President of Burritt's operation for the manufacture of millwork and sales of lumber.

The patent describes a thin flat metal level (preferably steel) with the vial cases attached to the outside surface, the idea being to reduce the overall weight of the tool. Two holes are to be provided to allow the tool to be attached to a long straight edge.

This patent appears to describe, exactly, a level produced by the Southington Cutlery Company of Southington, CT. The Southington level is not marked as being patented, and could possibly have been produced for a time after the patent protection expired. If so, then the level was produced for a short period between 1905 and 1908. The fact that the level utilized lined vials would indicate that production did not start until some time after the patent had expired because lined vials were not generally available until sometime after 1900.

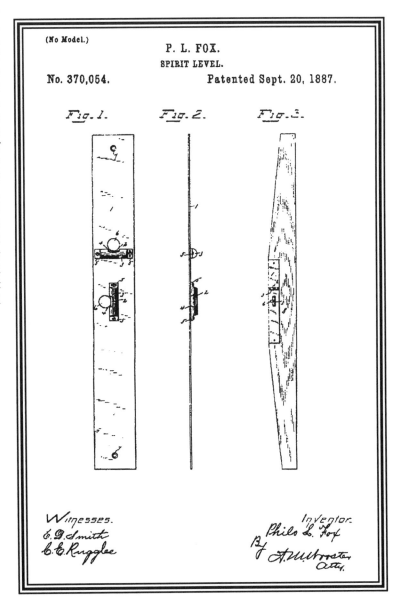

(No Model.)

P. L. FOX.
SPIRIT LEVEL.

No. 370,054.

Patented Sept. 20, 1887.

Fig. 1.   Fig. 2.   Fig. 3.

Witnesses.
C. D. Smith
C. E. Ruggles

Inventor.
Philo L. Fox
By A. M. Wooster
Atty.

SPIRIT LEVEL STOCK

For information on Edwin A.and Charles Stratton, see their patent of March 1, 1870 on page 30.

The patent describes several ways to accomplish the relationship between the pins, the strips and the end plates. These include (1) turning a round end on the strips, using a pin which could be driven into the strip from the end and peened over, (2) using a pin which could be soldered into place, and (3) drilling and threading the ends of the strip to receive a screw, which would be inserted from the outside of the end plate.

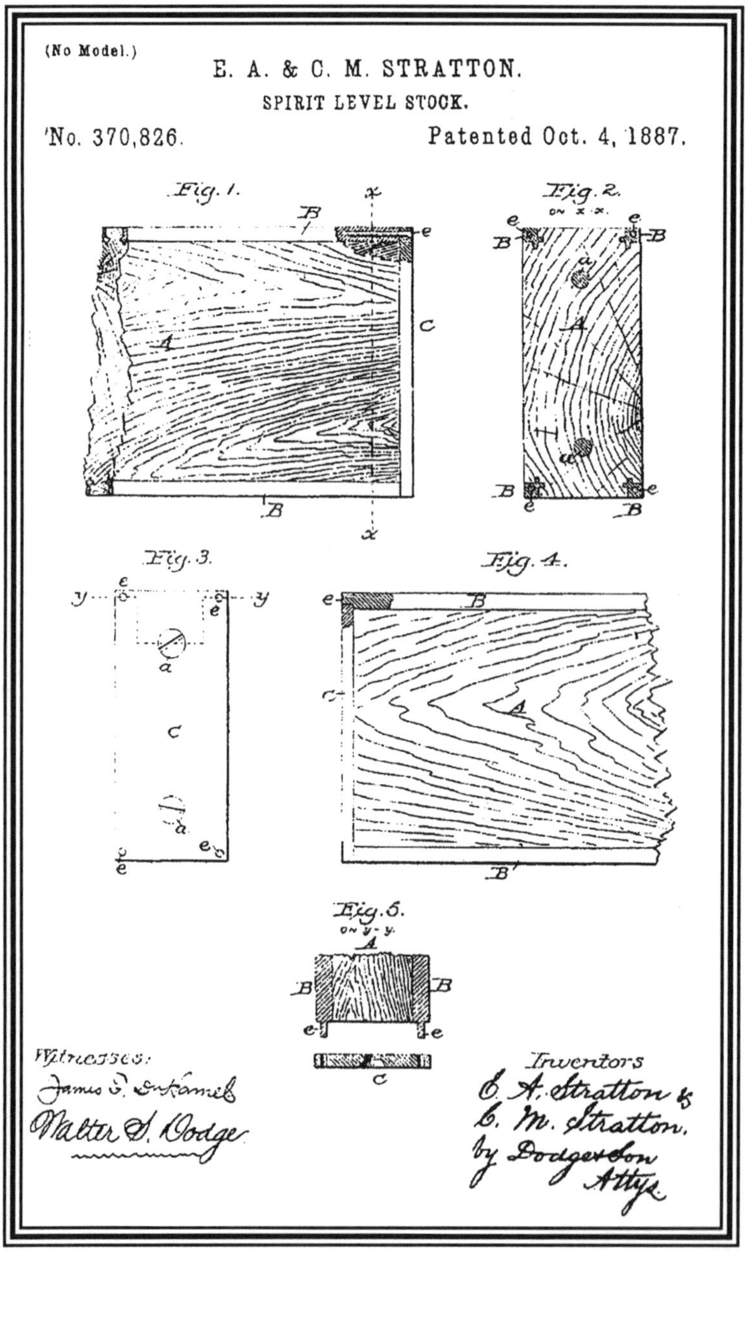

This patent concerns extending the pins through the brass end caps. Also the patent prescribes painting or saturating the ends of the level stock with a waterproof cement. These techniques were apparently used before they were patented because they were found on most Stratton Bros. levels.

(No Model.)

E. A. & C. M. STRATTON.
SPIRIT LEVEL STOCK.
No. 370,826.   Patented Oct. 4, 1887.

Fig. 1. Fig. 2. Fig. 3. Fig. 4. Fig. 5.

Witnesses:
James S. Infames
Walter S. Dodge

Inventors
E. A. Stratton &
C. M. Stratton.
by Dodgerson
Attys

PLUMB LEVEL FOR BORING BITS

George Woods was an upholsterer,[19] who maintained a listing in the Needham Business Directory. He had a house on the corner of Great Plain and Warren Streets. By 1893, his listing showed no occupation and he was no longer listed after1910.

This patent describes a tool consisting of a V-shaped hanger that fits over the shank of the bit; a U-shaped pendulum pivoting from the lower ends of the hanger; and an indicator arm attached to the pendulum, whose position can be read from a scale on the top portion of the hanger, i.e., an inclinometer.

The large open portion of the U is said to be designed to allow the level to be slipped over the bit.

The patentee also claims that the level can be used as an ordinary inclinometer by placing the side, c'c', against the side of the surface to be investigated.

This is a simple little device meant to be suspended from a boring bit being used in a carpenter's brace. No such tool has been identified, to date, and it seems impractical for boring vertical holes in spite of casual directions for its use in that manner.

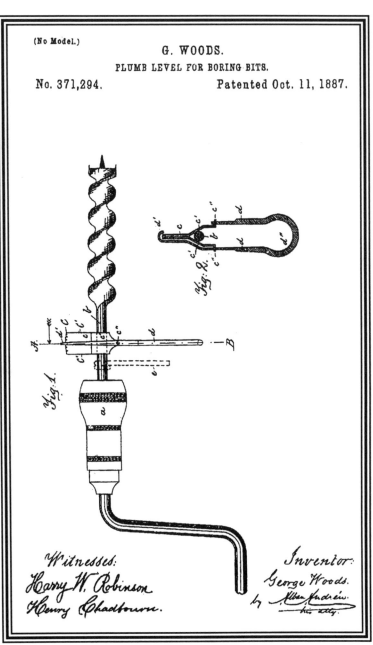

(No Model.)

G. WOODS.
PLUMB LEVEL FOR BORING BITS.

No. 371,294.   Patented Oct. 11, 1887.

Witnesses:
Harry W. Robinson
Henry Chadbourne.

Inventor:
George Woods.
by Allen Andrew
his atty.

---

[19]   *History and directory of Needham, Mass for 1888-89* by A. E. Foss and Co.

SPIRIT-LEVEL

There were two Charles B. Longs in the Worcester directory about the time of this patent. Both were machinists. One first appears in 1883, while the other has been in the directories since at least 1859. The patent will be attributed to the later C. B. Long. A Charles B. Long, probably the father, was listed in the New England Business Directory of 1856 as a machinist in Manchester, NH. He was not listed in 1860.

The patent only deals with a rotating cover for the spirit vial. No consideration was ever given to the design of the stock. i.e., the patent does not specify an inclinometer or even a metal stock.

Examples of Long's patent levels are known, and all are inclinometers that must be manually set.

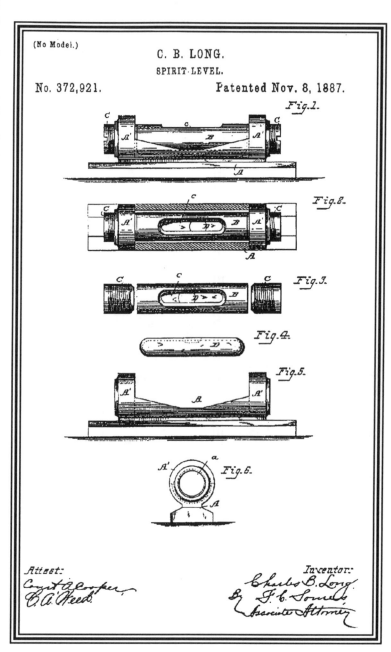

(No Model.)

C. B. LONG.
SPIRIT-LEVEL.

No. 372,921. Patented Nov. 8, 1887.

Fig.1.

Fig.2.

Fig.3.

Fig.4.

Fig.5.

Fig.6.

Attest:
Inventor:
Charles B. Long

Almeron W. Wickham and James M. Roach
Burnside, Connecticut

November 22, 1887
373,627

LEVELING-INSTRUMENT

At the time of this patent, Almeron Wickham was working in a paper mill along with another member of his family.   Nothing is known about the job or education background of Roach.

There was no listing in the Hartford City Directory for co-patentee James M. Roach. The only James Roach found in any Hartford City Directory during the period was a blacksmith listed in 1876.

The patent describes a combination tool, i.e., an inclinometer, sighting level, ruler, dividers, and compass with a pitch scale attached. The spirit vial (shown at *11*) is curved to 90° and set in the end of the piece. The sighting tube is obvious in *Fig.2*. The rule is graduated as shown in *Figs.4 and 5*. The pivot pin *15* is meant to be used in the hole, a, to support the tool when measuring angles.

The patent suggests that the instrument would be useful for determining the height of objects at a distance using the sighting level and the trigonometric scales on the side of the level.

No examples of the tool envisioned by this patent are known to date.

PATENT DRAWINGS ON NEXT PAGE.

A. W. WICKHAM & J. M. ROACH.

LEVELING INSTRUMENT.

No. 373,627.

Patented Nov. 22, 1887.

2 Sheets—Sheet 1.

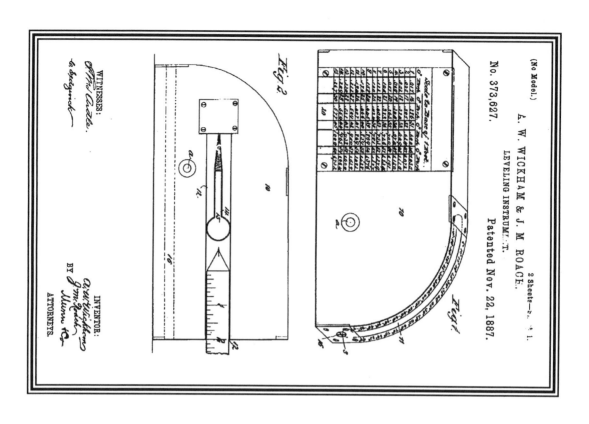

Fig. 1

Fig. 2

A. W. WICKHAM & J. M. ROACH.

LEVELING INSTRUMENT.

No. 373,627.

Patented Nov. 22, 1887.

2 Sheets—Sheet 2.

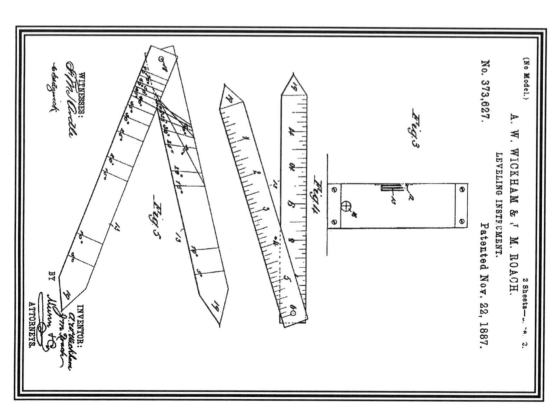

Fig. 3

Fig. 4

Fig. 5

Edwin A. Stratton and Charles M. Stratton
Greenfield, Massachusetts

May 22, 1888
383,196

SPIRIT-LEVEL

For information on Edwin A. and Charles Stratton, see their patent of March 1, 1870 on page 30.

According to the patent, one claim is that the bead is not soldered on the wire and thus shall be moveable with a push. An alternative idea is that the bead will be fixed on the wire, but the wire will not be fixed to the vial cartridge and thus can be slid in either direction. (Oxidation of metal surfaces may make movement very difficult, however.)

The elements of this patent were utilized on the plumb fixture of most Stratton Bros. levels. This patent is for the moveable bead used to indicate the true level or plumb position of the bubble.

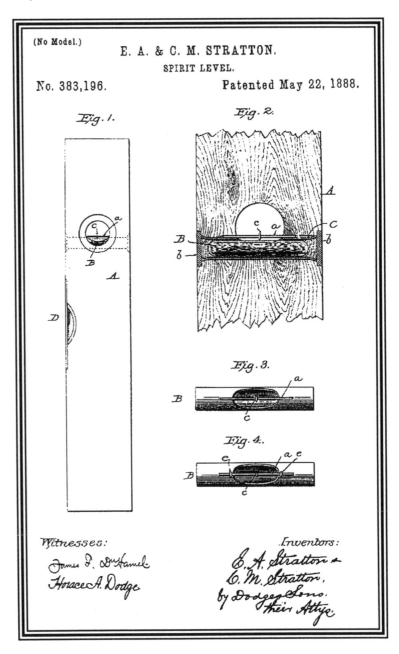

(No Model.)

E. A. & C. M. STRATTON.

SPIRIT LEVEL.

No. 383,196.   Patented May 22, 1888.

Fig. 1.

Fig. 2.

Fig. 3.

Fig. 4.

Witnesses:
James I. DuHamel
Horace A. Dodge

Inventors:
E. A. Stratton &
C. M. Stratton,
by Dodge Sons,
their Attys.

James C. Hutton
Corvallis, Oregon

July 3, 1888
385,516

Assigned one-half to Patrick J. McElroy of East Cambridge, MA

SPIRIT-LEVEL

Patrick J. McElroy was a glass blower by trade.

No information is available about James C. Hutton at this time.

The proposed device incorporates a 90° segment-shaped spirit vial in one or both ends of the level stock. The unit containing the spirit vial is made of two sections, one of which is fixed to the basic stock. The spirit vial is inserted into an annular depression in the fixed half and, after proper orientation, is embedded in plaster. Additional plaster is placed on the exposed side of the vial and a mating piece is attached to the fixed portion of the stock. The edge of the stock is calibrated such that angles between 0° and 90° can be read. Vials may be truly circular or contain only a quarter segment of a true circle.

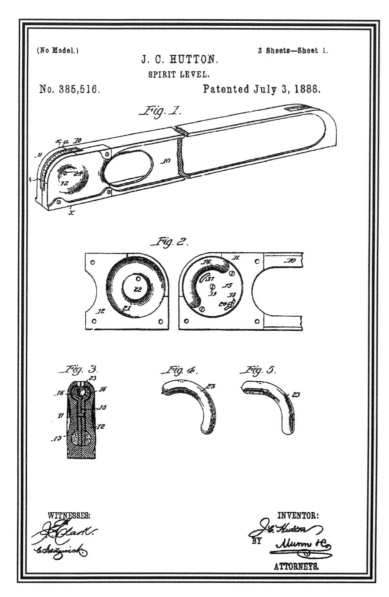

Another version of the device, as shown in *Fig.6* on the next page, has a mantel clock shape and utilizes a semicircular segment for a spirit vial. Here the vial can again be truly circular or can contain only one half of a true circle. In this version, the scale, similarly provided around the top edge of the mantel clock, can be read +/- 90° from the top dead center.

A third proposed version of this device utilizes much smaller quarter circle segments put into the upper corner of each end of the level. In this fashion, the remainder of the corner can be used for plates to protect the level against damage.

No such device has been identified to date.

J. C. HUTTON.

SPIRIT LEVEL.

No. 385,516.

Patented July 3, 1888.

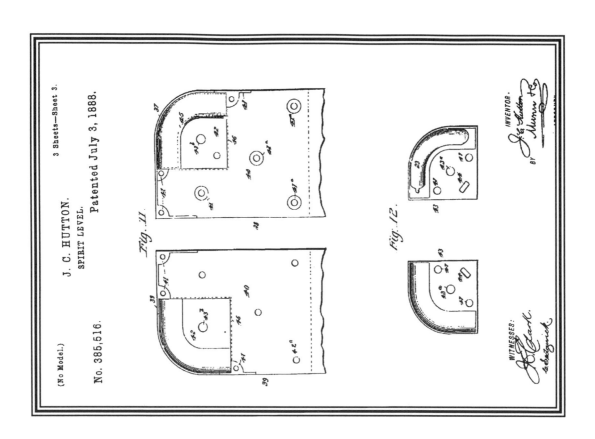

Fig. 11.

Fig. 12.

WITNESSES:

INVENTOR.

BY

J. C. HUTTON.

SPIRIT LEVEL.

No. 385,516.

Patented July 3, 1888.

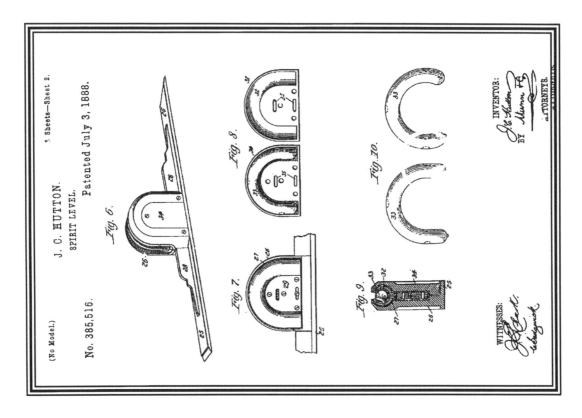

Fig. 6.

Fig. 7.

Fig. 8.

Fig. 9.

Fig. 10.

WITNESSES:

INVENTOR:

BY

ATTORNEYS.

Assigned to the Stanley Rule and Level Company of New Britain, CT

ATTACHMENT FOR CARPENTERS' RULES

G. F. Hall, the co-patentee, was from New Jersey. Hall held the first patent on the Odd-Jobs (January 25, 1887). He also held other patents for similar tools. He will be discussed in detail in Volume II.

For information on Justus A. Traut, see his patent of October 6, 1868 on page 28.

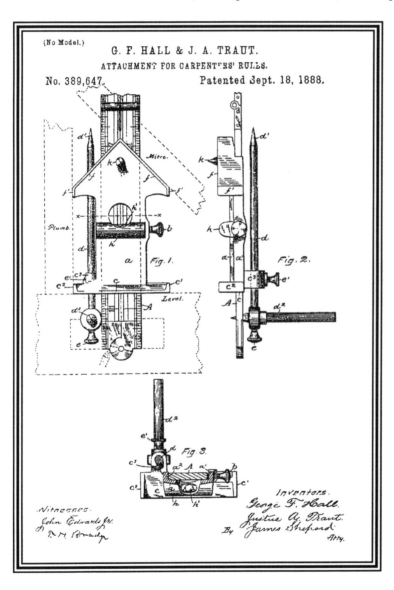

(No Model.)

G. F. HALL & J. A. TRAUT.

ATTACHMENT FOR CARPENTERS' RULES.

No. 389,647. Patented Sept. 18, 1888.

Fig. 1. Fig. 2. Fig. 3.

Witnesses.
John Edwards Jr.
R. H. Brady

Inventors.
George F. Hall
Justus A. Traut
By James Shepard
Atty.

While various claims are made for what is new in this patent, the primary difference between tools made by the first, and the combination of the first and second, patents is found in the position of the receptacle for the scribe. Other less obvious differences are found in certain proportions and in the use of vertical side ribs extending both in front and rear of the body.

This is the second patent for the Odd-Jobs, and the existing tools are faithful to the concepts of the patent.

Albert D. Goodell
Millers Falls, Massachusetts

October 16, 1888
391,242

Assigned to the Millers Falls Co. of Millers Falls, MA

SPIRIT-LEVEL

For information on Albert Goodell, see his patent of December 30, 1884 on page 67.

The patent describes an adjustment for the plumb fixture and for each end of the level fixture. On the level fixture, the first piece, a threaded post on a base is secured to the bottom of the mortise in the level stock. (The threaded post appears to be similar to a machine screw with a round slotted head, like a wood screw; there is a slot in the other end of the screw that would accept a screw driver.) A spring is placed over the post followed by one end of the vial casket and a captive nut. The nut is kept captive by a square mortise in the end of the vial casket. The slot in the top of the post allows the post to be turned, and a recess beneath the head provides room for the adjusting screw to move down.

The plumb adjustment is similar except the secured portion (as shown in *Fig.4*) is opposite the adjustable post and making an adjustment causes the plumb vial case to tilt.

In the only example known to the author, the post has been replaced by a square shouldered machine screw and a plate fixed to the bottom of each mortise had threaded holes. The machine screw was fitted with an elliptical head having the long sides flatted. The holes remain in the top plates but the plates must be removed to make adjustments.

This patent was for an adjustable wooden level. It appears that it may have been used for some short period of time although in some modified form.

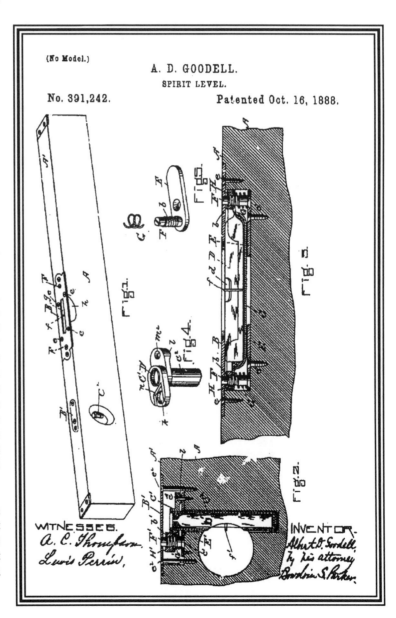

(No Model.)

A. D. GOODELL.
SPIRIT LEVEL.

No. 391,242.

Patented Oct. 16, 1888.

WITNESSES.
A. C. Thompson.
Lewis Perrin,

INVENTOR.
Albert D. Goodell,
by his attorney
Bordsin S. Parker

Justus A. Traut
New Britain, Connecticut

May 7, 1889
402,869

Assigned to the Stanley Rule and Level Company of New Britain, CT

SIGHTING ATTACHMENT FOR LEVELS

For information on Justus A. Traut, see his patent of October 6, 1868 on page 28.

This was the patent for the № 1 sights for wooden levels and known sights are faithful to the design proposed in the patent.

Although the patent drawing shows an adjustment screw on the right hand sight, the use of such a screw on one or both sight pieces was said to be optional. As shown by the drawing, the front sight had a peep hole and the back sight had a single horizontal cross wire. The manufactured product had crossed wires.

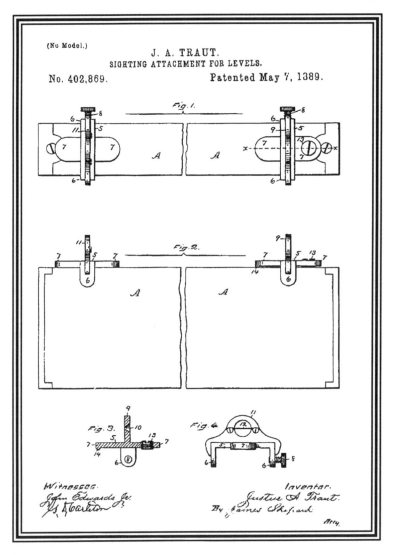

(No Model.)

J. A. TRAUT.
SIGHTING ATTACHMENT FOR LEVELS.

No. 402,869.

Patented May 7, 1389.

According to the patent, the rear sight *could* be fitted with a screw for adjusting its height above the level stock, in order to put both sights in the same horizontal plane.

Justus A. Traut
New Britain, Connecticut

June 18, 1889
405,624

Assigned to the Stanley Rule and Level Company of New Britain, CT

SIGHTING ATTACHMENT FOR SPIRIT LEVELS

For information on Justus A. Traut, see his patent of October 6, 1868 on page 28.

This patent is for sights that were inserted into a receptacle in the top side of the stock of wooden levels. The level stock was to be fitted with a threaded insert at each end. Each piece of the sights was to be attached to the threaded insert with a screw when needed.

No sights made according to the specifications of this patent have been found to date.

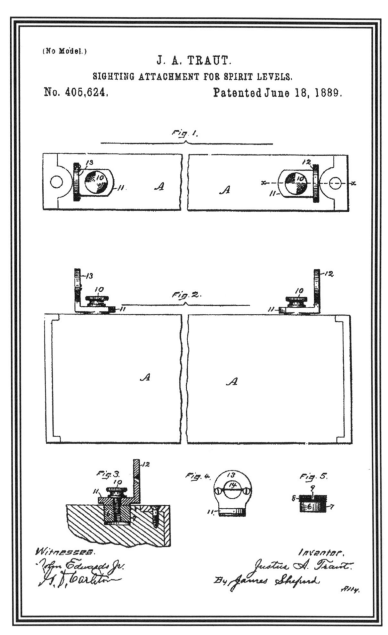

(No Model.)

J. A. TRAUT.
SIGHTING ATTACHMENT FOR SPIRIT LEVELS.
No. 405,624.   Patented June 18, 1889.

DEVICE FOR MEASURING THE HEIGHT OF HORSES

Harry Lowe was in the saddle goods business in 1887-88. In 1891 and 1892, he was listed as being in the manufacture of horse specialties and, in 1895, as being in the manufacture of horse goods and ladders.

This patent describes a long cylindrical device containing a tape measure on one end. The tape was wound on a shaft and maintained in that position by a spring. It could be pulled down by attaching a heavy plumb bob weight to the end of the tape. The long cylinder contained a spirit vial. To operate the device, the cylinder is placed on the horse, the weight is attached to the tape and the tape is allowed to unreel from shaft. The device is grasped at *I*, which secures the tape, and the cylinder is raised slightly to allow the weight to pull the tape into a plumb position. The cylinder is placed back on the horse and the securer *E* is released, which allows the spring to pull the tape taut again. The height is then read from the tape.

The device was manufactured and is quite faithful to the patent specifications. Harry Lowe could have been the manufacturer.

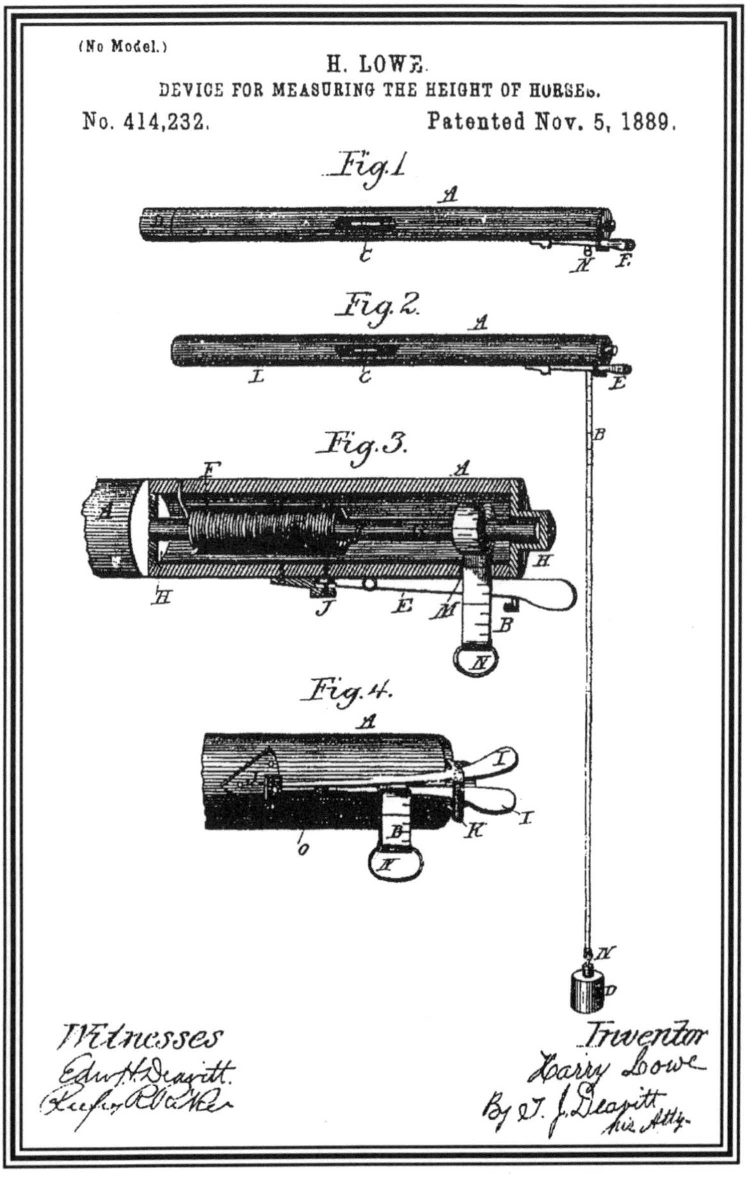

(No Model.)

H. LOWE.
DEVICE FOR MEASURING THE HEIGHT OF HORSE.
No. 414,232. Patented Nov. 5, 1889.

Fig.1.

Fig.2.

Fig.3.

Fig.4.

Witnesses
Edw. H. Deavitt.
Rufus R. Blake.

Inventor
Harry Lowe
By T. J. Deavitt,
his Atty.

Joseph McLaughlin
Erving, Massachusetts

SPIRIT-LEVEL FOR BIT-BRACES

Joseph McLaughlin was an employee of Millers Falls Company. He was known to have lived in Erving during 1887-88 and in Millers Falls in 1893.

The patent drawings are self-explanatory. A surface type or bulls-eye level is set into the top of the head of a brace, and a plumb vial is set into the head on a vertical surface on the side. Thus, the user can determine horizontal and vertical directions for driving the bit.

No such device has been identified to date.

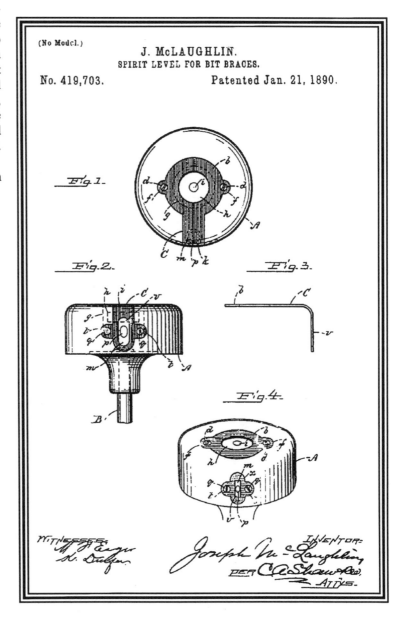

Justus A. Traut **February 18, 1890**
New Britain, Connecticut **421,786**

Assigned to the Stanley Rule and Level Company of New Britain, CT

SPIRIT LEVEL

For information on Justus A. Traut, see his patent of October 6, 1868 on page 28.

There were two major features of this patent: (a) adjustment of top vial and (b) indentures in the center of the top plate. The top, or level, vial was to be adjustable from the top of the top plate using one screw inserted into one end of the vial cartridge. The other end of the vial cartridge was inserted into a slot in a brass piece extending below the top plate. Springs between the vial cartridge and the top plate were used on both ends. The inden-tures, or notches, in the top plate were used to indicate the center of the vial, and to help the user recognize a level condition. They also give a reference point for adjusting.

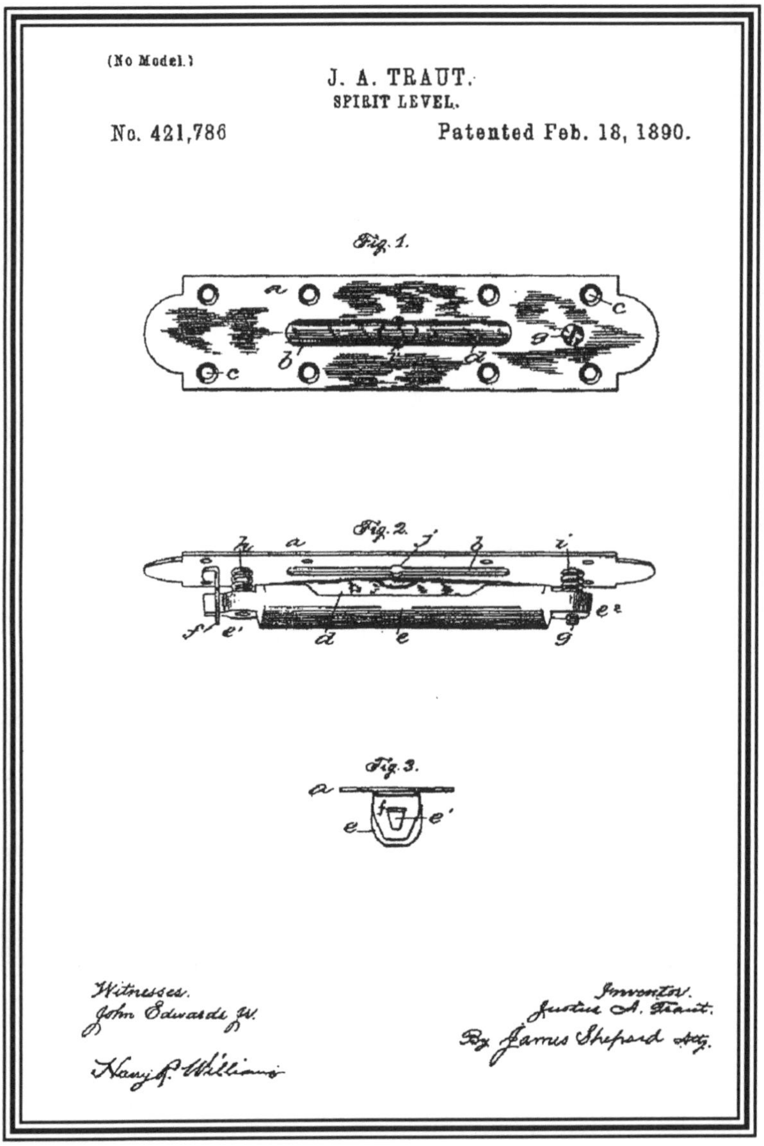

(No Model.)
J. A. TRAUT.
SPIRIT LEVEL.
No. 421,786
Patented Feb. 18, 1890.
Fig. 1.
Fig. 2.
Fig. 3.
Witnesses.
John Edwards Jr.
Harry R. Williams
Inventor.
Justus A. Traut.
By James Shepard Atty.

This patent was employed in essentially all standard format Stanley Rule & Level Company carp-enters' levels beginning in 1890. The manufactured products were faithful to the concept of the patent.

The patent does not men-tion any adjustment feature for the plumb vial. How-ever, some levels of the period were fitted with a cover piece to protect the adjustment screws. The cover piece fit into small slots on either side of the adjustment screws. These covers were supplied for both the plumb and the level vial adjusting screws, and both were marked with the Feb. 18, 1890 patent date. The covers were apparently lost quite easily, and are usually missing.

William Martin
Salem, Massachusetts

March 19, 1890
423,484

INCLINOMETER

William Martin, a carpenter by trade, had a home at 4 Herbert Street. He was an independent business man with offices at 64 Lafayette Street in Salem. His listing in the business directory for 1899-1900 described him as carpenter and inventor. In 1904 he was a partner in Abbott and Martin, Carpenters and Builders. In 1915, the firm was also listed in the business directory under Contractors. Martin died in April 1918.

The patent provides for the following: To a wooden level stock containing both plumb and level spirit vials, is added a provision to adjust one end of the stock. The inclinometer function of the apparatus is accomplished utilizing a hollow post containing a rod with a foot (shown in *Fig.3*). The inclinometer rod is provided with a spring at the bottom to force the foot down. The rod carries an index pointer and the position of the inclinometer is determined from a scale on a slot in the post. A thumb screw on the side of the inclinometer portion locks the rod in the desired position.

For storage, the entire apparatus can be rotated 90° into a cavity provided for it in the stock. A portion of the split stock then is folded back into a covering position.

No apparatus of this description has been identified to date.

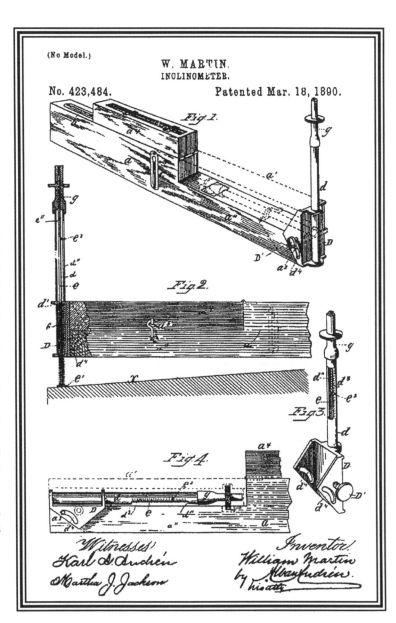

(No Model.)

W. MARTIN.
INCLINOMETER.

No. 423,484.   Patented Mar. 18, 1890.

Fig. 1.

Fig. 2.

Fig. 3.

Fig. 4.

Witnesses
Karl A. Andrén
Martha J. Jackson

Inventor
William Martin
by Andrén
his atty

Justus A. Traut
New Britain, Connecticut

March 25, 1890
423,969

Assigned to the Stanley Rule and Level Company

SPIRIT LEVEL

For information on Justus A. Traut, see his patent of October 6, 1868 on page 28.

The levels, using this patent, are recognizable by the wide, flush-mounted porthole surrounds described in the patent. The porthole surrounds were really inserts carrying the spirit vials and lined the inside of the port hole as well as protecting the surrounding surface. Another feature of the patent, however, is the cast iron fixture behind the porthole surround. This fixture rested on another cast iron piece that was firmly fixed to the level stock. The uppermost piece is fitted with two slots through which the brass porthole surround is attached with screws. This piece can rotate for adjustment of the spirit vials.

This patent was employed by Stanley on the № 25, 30, 50 and 100 levels. The patent features were faithfully followed in the products.

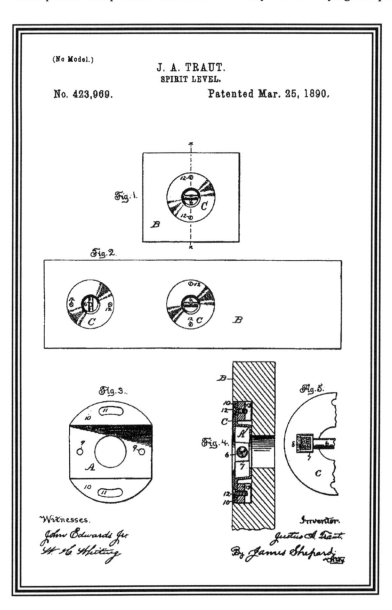

(No Model.)

J. A. TRAUT.
SPIRIT LEVEL.

No. 423,969.   Patented Mar. 25, 1890.

Fig. 1.

Fig. 2.

Fig. 3.

Fig. 4.

Fig. 5.

Witnesses.
John Edwards Jr
H. H. Whiting

Inventor.
Justus A. Traut
By James Shepard

SPIRIT-LEVEL

For information on Philo L. Fox, see his patent of September 30, 1887 on page 83.

This patent was for an lightweight level consisting of two wooden rails connected by four I-shaped posts. The level vial is readable from below via two long narrow slits extending down from the vial to the outside edges of the top rail. The patent specifies that all the vials are to be contained in transparent tubes. As a result, the plumb vials are readable from any direction because there is no metal to block the view.

A level has been produced according to this patent. It is unmarked, except for the patent date, and the manufacturer is unknown.

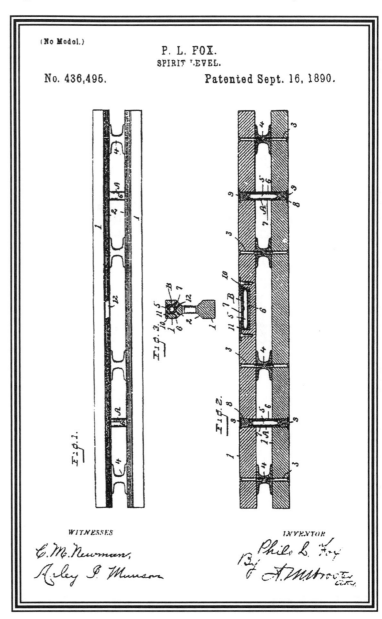

(No Model.)

P. L. FOX.
SPIRIT LEVEL.

No. 436,495.

Patented Sept. 16, 1890.

WITNESSES

INVENTOR

SPIRIT LEVEL

Henry Green was a pattern maker in Hartford in 1890-92. This was Green's first patent. For that time the application took longer than average to be approved; it had been filed on March 27, 1889.

The patent is for an inclinometer with a semicircular bubble tube and a moveable arm to frame the bubble and highlight the scale. The face of the inclinometer is graduated and the moveable arm is provided with a slit (*Fig.4* at *g*) through which the scale can be read precisely.

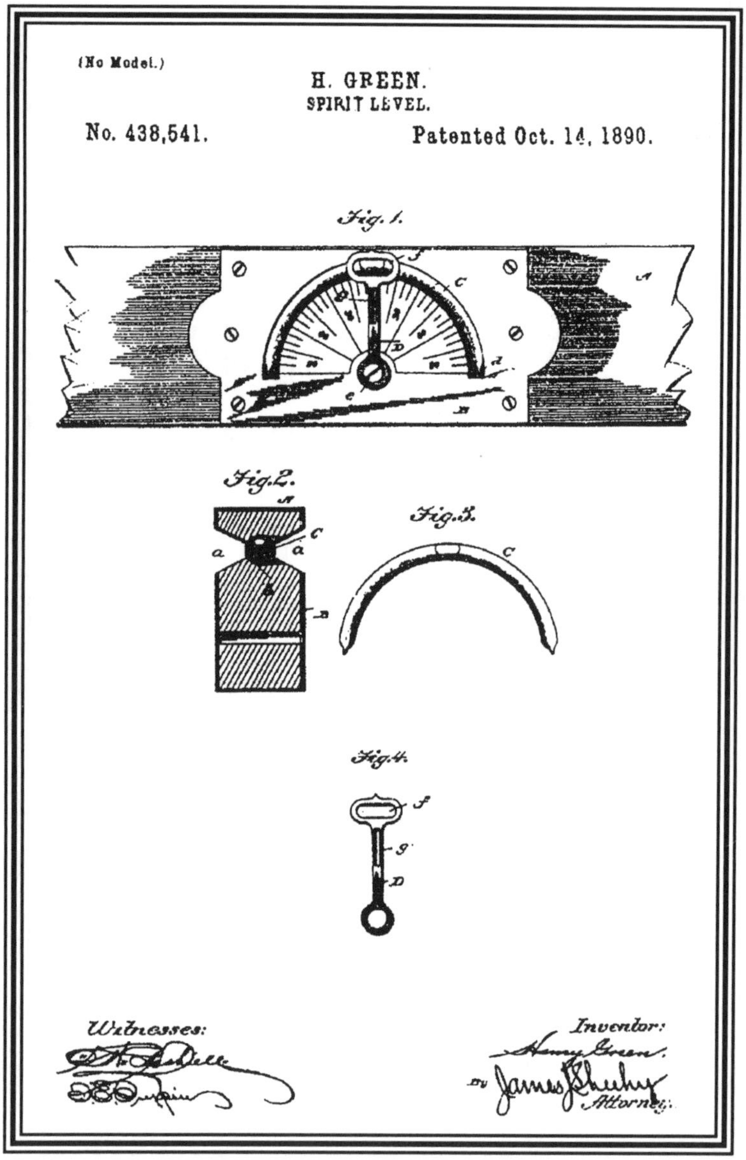

The moveable arm is not meant to be free swinging and can be positioned to a pre-set angle, and the level inclined until the bubble is centered.

There are no known examples of levels made according to this patent.

(No Model.)

H. GREEN.
SPIRIT LEVEL.

No. 438,541.

Patented Oct. 14, 1890.

Fig. 1.

Fig. 2.

Fig. 3.

Fig. 4.

Witnesses:

Inventor:
Henry Green
by James Shelhy
Attorney.

Henry Green
Hartford, Connecticut

March 31, 1891
449,609

Assigned to John J. Tower of New York, NY

SPIRIT-LEVEL

For information on Henry Green, see his patent of October 10, 1890 on page 104.

The patent envisions a spirit vial in a cartridge that is fixed to a revolving ring. A 360^0 scale is placed on a fixed inner disk. The back of the tool is fitted with a notch and a lock screw to tighten the device on a round shaft such as a drill bit. It also contains a flat surface that could be clamped to a rule or a square using the lock screw.

This, Henry Green's second patent, was used on the Tower & Lyon rule and square level. Examples are known, both with the split ring shown in the patent drawing, and with a solid ring. Also, both brass and nickel plated types are known.

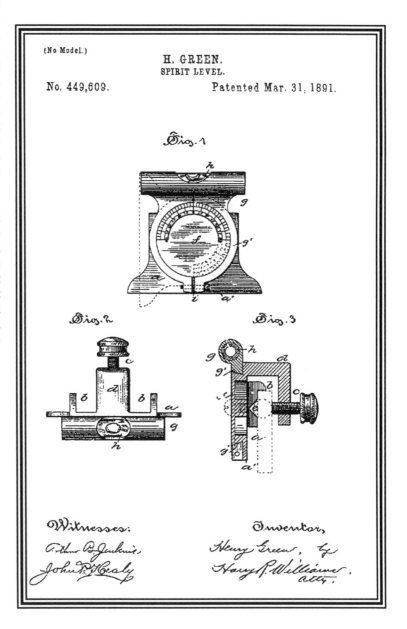

(No Model.)

H. GREEN.
SPIRIT LEVEL.

No. 449,609.

Patented Mar. 31, 1891.

Henry Green
Hartford, Connecticut

April 14, 1891
450,457

Assigned to the Acme Rule Company of Salisbury, CT.

SPIRIT-LEVEL

For information on Henry Green, see his patent of October 10, 1890 on page 104.

Acme Rule Co. was a maker of patented inclinometers as well as rules and novelty items.

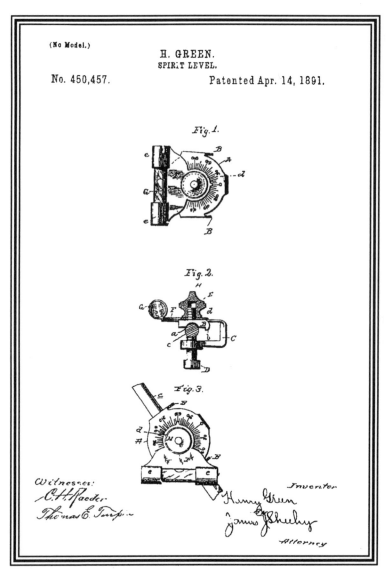

(No Model.)

H. GREEN.
SPIRIT LEVEL.

No. 450,457.   Patented Apr. 14, 1891.

Fig. 1.

Fig. 2.

Fig. 3.

Witnesses:
C. H. Raeder
Thomas E. Turpie

Inventor
Henry Green
James J. Sheehy
Attorney

The level described by this patent differs from that of Green's preceding patent in that the wing carrying the spirit vial is fixed to the vertical post in the center of the tool. When the spirit vial is turned, a pointer on the lower part of the wing indicates the angle against the scale on the fixed body of the tool. This tool is meant only for boring bits.

This, Henry Green's third patent, was similar to 449,609, but the tool was meant as an inclinometer level for bits. To date, there are no known examples of this tool.

Justus A. Traut
New Britain, Connecticut

June 2, 1891
453,452

Assigned to the Stanley Rule and Level Company of New Britain, CT

SPIRIT LEVEL

For information on Justus A. Traut, see his patent of October 6, 1868 on page 28.

The patent specifically claims a groove in a level stock that extends into the porthole. An additional claim was for a grasping surface. However, Traut claimed the purpose of the invention was "to provide a convenient means for handling", "to improve the appearance" and "to make the stock less liable to warp out of shape."

This is a patent is for what became known as the hand-y groove (or grip). First products were faithful to the design shown in the patent drawings and claimed in the patent.

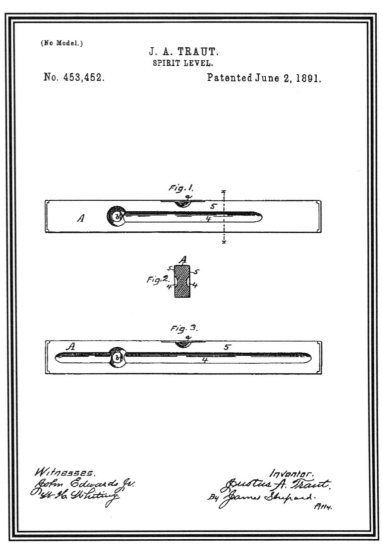

SQUARE FOR LEVELING AND ALIGNING SHAFTING

No information about Moses Oliver is available at this time[20].

The patent describes a three piece tool. A holder bar f in the drawing carries a spirit level at any convenient point along its length. At one end of the holder bar there is a square, and the vertex of the square is located beyond the end of the bar. The square can be adjusted on the holder bar to either of the two positions shown in *Fig.2*. The position shown by the solid line in *Fig.2* is used for leveling the shaft; while the position shown by the dotted line is used for aligning the shaft. Most of the material in *Fig.1* is instructional for use of the tool.

The tool is known to have been produced; however, details of its construction are not known at this time. The author has not personally observed the tool.

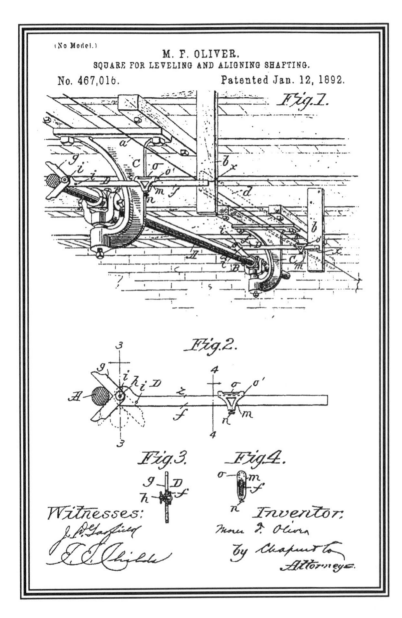

20   This patent was discovered too late to research Moses Oliver or to seek information about the tool and likewise, the existence of the tool was learned too late to be able to observe its detail.

William Praddex
Lawrence, Massachusetts

February 23, 1892
469,451

SPIRIT-LEVEL

William Praddex was a painter of railroad cars and carriages and, in 1892, he worked for the Boston and Maine Railroad. He was listed in the Lawrence City Directories from 1868 to 1901. In 1901, he was an assistant foreman

There are three features to this patent: (1) using a stock that incorporates a groove running horizontally from end to end of the level; 2) cutting away portions of the top plate along the edges for the length of the vial (to prevent or reduce the instance of the top plate striking something if the level should fall over); and (3) wrapping a thin wire around the vial to indicate the level position.

The patent seems to have been inspired by Traut's Hand-y groove patent of June 2, 1891. Closer examination shows it to have been filed (April 11, 1891) after Traut's filing, but prior to the granting of Traut's patent Nº 453,452. No example of the Praddex device has been identified to date.

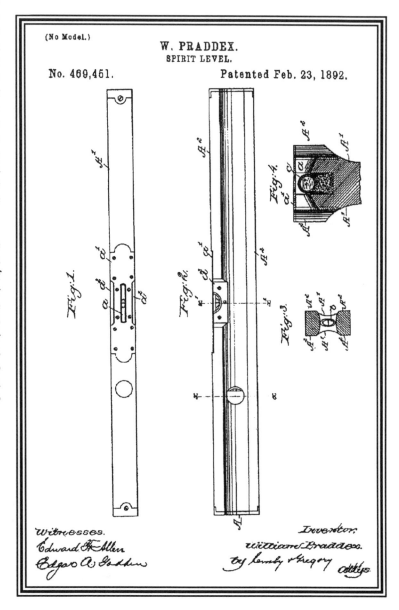

(No Model.)

W. PRADDEX.
SPIRIT LEVEL.

No. 469,451.

Patented Feb. 23, 1892.

PLUMB-LEVEL

Augustus Taylor was not listed in the Claremont City Directories for 1881, 1887, 1893-4, 1896-7 or 1899, nor was he listed in the 1880 or 1900 Census. The witnesses on this patent were Harvey Lincoln and Arabella his wife. Harvey was a foreman at the Maynard Shoe Company in Claremont. Their relationship to Taylor is unknown.

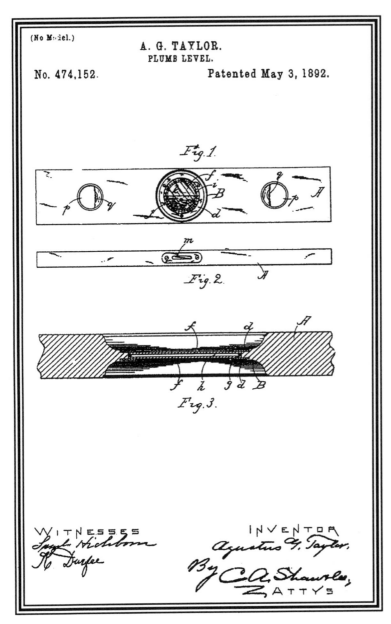

(No Model.)

A. G. TAYLOR.
PLUMB LEVEL.

No. 474,152.   Patented May 3, 1892.

Fig. 1.

Fig. 2.

Fig. 3.

WITNESSES   INVENTOR
                             Augustus G. Taylor,
                             By C. A. Shawles,
                             ATTYS

This device was envisioned as a standard double plumb level, fitted with a circular chamber that could be viewed from either side. The chamber was to be air tight and filled with a light liquid, preferably ether (because of the negligible effect of temperature on the expansion of the liquid). The outer edge of the chamber was graduated so that the device could also be used as a protractor.

To date, no example of this device has been identified.

LEVEL HANGER FOR SHAFTING

Nathaniel D. Poole was a carpenter by trade according to his listing in the city directory. He died sometime prior to 1909.

The application of this device allows an ordinary level to be used when hanging shafting, and allows the work to be done from the shop floor. The patent drawing is really self-explanatory. The device consists of two pieces. The upper piece hooks over the shaft, has a hollow, internally threaded, ruled post, and fits inside the lower piece. The second, or lower, piece contains a hanger at the bottom to hold the level; it is of sufficient diameter to fit over the top piece, and at the base is fitted with a long thumb screw that can engage the threads in the top piece. The use of two or more of these on a shaft, with a carpenter's level in the bottom hanger, allows measurement or setting of the slope of the shaft.

No such device has been located for study.

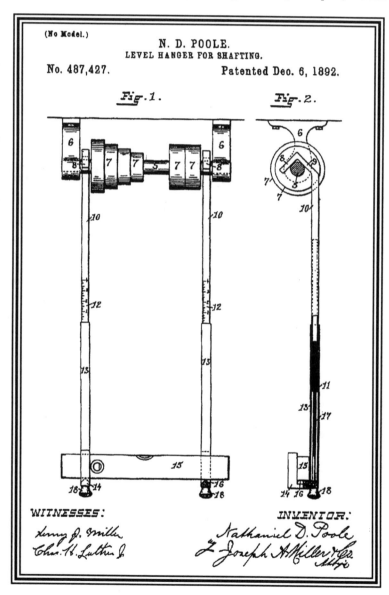

SPIRIT LEVEL

For information on Justus A. Traut, see his patent of October 6, 1868 on page 28.

This patent includes several features. One of these features is for a base with opposing abutments, such that a level case can be fastened between them (not on top, as later № 34 levels were made.) A second feature is that the level case is to be fastened between the abutments, with screws through the abutments and into the ends of the level case, in a direction parallel to the level case, with one end being adjustable for a short distance. Two types of protective devices are described in the patent. The first device incorporates a pair of cylindrical covers, each of which is about one-third of the length of the level case. In this variation, the protection is invoked by *sliding* one of the one of the covers from the end to the center to cover the exposed portion of the vial. The second type of protective device is for a *rotating* cover of approximately the same length as the level case. Note that according to the patent drawing, the protective cases have slots cut in them near the outer ends. These are to be creased in such a way as to offer a slight resistance to sliding and thereby remain where they were placed.

It seems noteworthy that this patent was not assigned to the Stanley Rule & Level Company. It does, however, appear to have been used by Stanley. Future patents by Traut were not again assigned to Stanley until Patent № 760,587 in 1904.

Both of these types of covers are known to have been used, although the sliding covers are extremely rare.

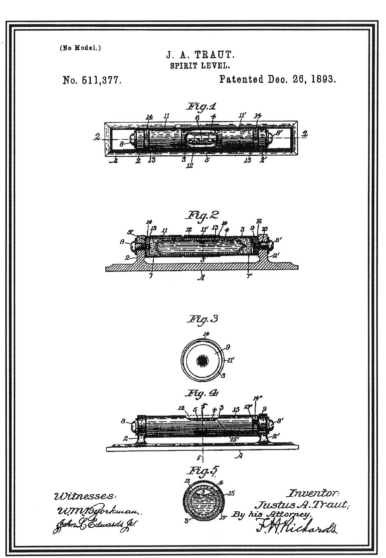

LEVEL

For information on Justus A. Traut, see his patent of October 6, 1868 on page 28.

This patent is for a plumb glass case with several incorporated features. The patent itself goes into great detail regarding the construction of the brass plumb case and the means by which it is held in the plumb recess of the stock. The case is held in the bottom of the plumb recess either by barbs on the end piece, which are set with a punch from the bottom side, or by a jam fit of the end piece without the barbs. The plumb is adjustable from the top end by a screw in an oblong hole. The patent also incorporates two center-marking points (N°s. *12* and *13* in *Fig.4*).

The features of this patent were used in many of the Stanley levels such as the N° 3.

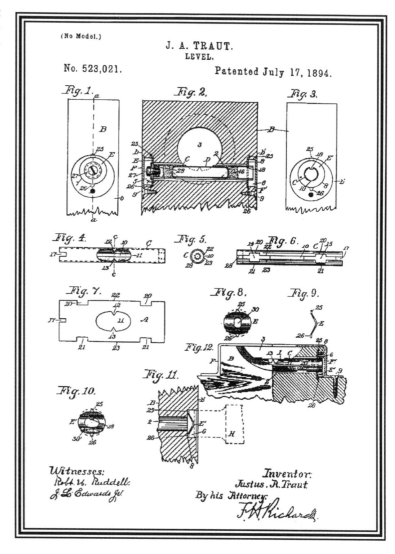

LEVEL

For information on Justus A. Traut, see his patent of October 6, 1868 on page 28.

This patent is for a cover plate or top plate with opposing points in the opening to mark the center.

The features of this patent were also used as shown in levels such as the Stanley № 3.

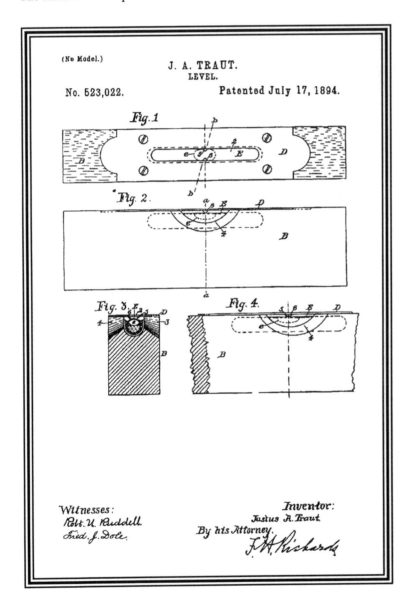

LEVEL

For information on Justus A. Traut, see his patent of October 6, 1868 on page 28.

The main features of the patent are a sheet metal blank formed into a vial casket, with tabs on the end for fastening to the stock. The blank also has tabs that fold over the ends of the vial. The sides of the casket are low enough to view the bubble through the side view. Springs are placed under the tabs and adjustment screws pass through them.

This patent for a simple vial casket was used in the Stanley Nºs. 60, 70, 80, 90 and others.

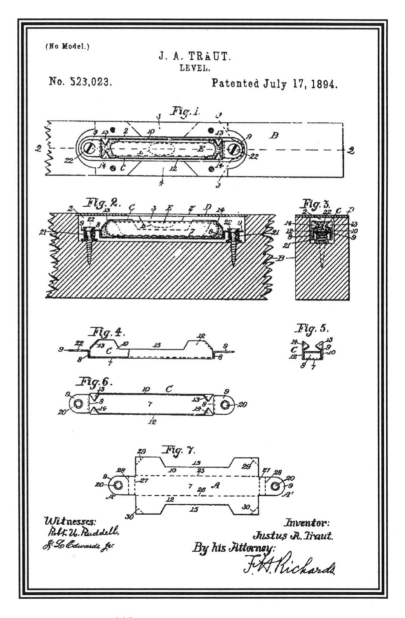

SPIRIT LEVEL

For information on Justus A. Traut, see his patent of October 6, 1868 on page 28.

This patent describes an adjustable bed or carrier. The tensioning feature of the patent includes notches in the carrier to prevent the wire from changing position. The tensioning wire, which goes around both the vial and the vial casket, also serves as an indicator of the center of the vial for adjusting purposes. Although not clearly shown in the drawings, there is an independent bottom plate which is secured to the bottom of the mortise. The vial carrier is, in turn, secured to the bottom plate with machine screws that can be used to adjust the position of the carrier.

Stanley levels made with features of this patent include the Nºs 96 and 98. They do not, however, include the tensioning wire shown as part number *11* in the figures.

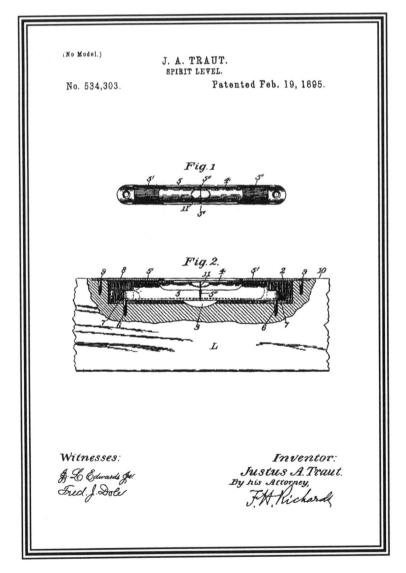

Carl A. Carlson
Orange, Massachusetts

June 18, 1895
541,151

BEVEL-SQUARE

Carl August Carlson was a citizen of Sweden, living in Orange. He was employed by the New Home Sewing Machine Company at the time of this invention and continued to be employed there for another 20 years.

This device is best described as a standard bevel with a provision for two removeable blades on the same end. The blades are attached to the stock with interlocking dovetail type grooves and moved using a geared wheel, which can be driven by a beveled gear attached to a knurled screw. The position of the removable blades may be changed to a fixed 45° bevel and a fixed 90° position. (Shown in *Figs.3* and *4* of this drawing.) With the fixed blades, the tool becomes a mitre square or a T-square respectively. A spirit level vial is inserted in one edge of the stock.

It seems that insertion of a spirit vial in a bevel square of some type or in the handle of a combination tool was quite common for purposes of patents. Most of these tools were, however, produced without the spirit vial. Thus one need not expect to find the vial in the manufactured tool nor should one consider a tool without the vial as automatically being a variation of that tool.

To date, no such bevel-square has been found in any configuration.

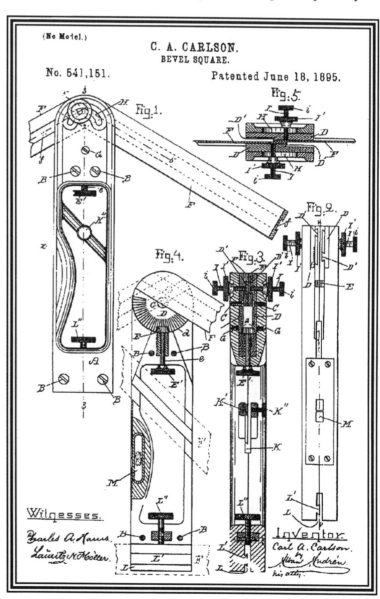

PLUMB-LEVEL

Lewis C. Raymond was a carpenter residing in Centre Rutland, according to Pelton's Directory of Rutland for 1887-88. In 1891-92, he was listed as being a farmer on West Creek Road in Center Rutland. In the Rutland City Directory of 1911, he was listed as "Louis C." retired.

This level incorporates two plumb-bobs. In order to indicate level, one consults a plumb-bob suspended from an external arch at the top center of the stock. The cord for the bob extends through a guide that incorporates a raised pointer-like section. In order to indicate plumb, one consults a plumb-bob suspended internally from the left end of the stock. This cord extends through a guide in a separate window that also incorporates a raised pointer-like section to indicate plumb when the cord is in line with the apex. Both bobs can be steadied by a lever extending through the stock and cupping the bob.

The pieces *J* shown in *Fig. 2* are covers for the holes *B* and *B'*.

No such device has yet been identified.

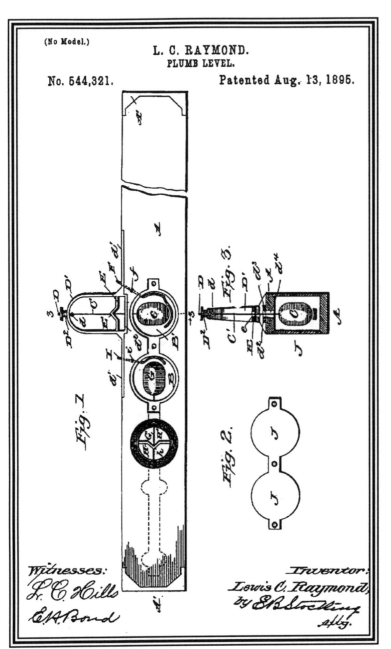

(No Model.)

L. C. RAYMOND.
PLUMB LEVEL.

No. 544,321.   Patented Aug. 13, 1895.

Witnesses:
L. C. Hills
E. H. Bond

Inventor:
Lewis C. Raymond,
by E. B. Stockling
atty.

DESIGN FOR A LEVEL FRAME OR BLOCK

For information on Justus A. Traut, see his patent of October 6, 1868 on page 28.

This is a design patent for a level frame. The claims of the design revolves around the concentric bands, the outer one being a brass ring, and the inner one being a beveled ring in the wood of the stock. The plumb bands are to be circular and the sideviews are to be semicircular.

This design was used for several levels, among them the Stanley Nºs 60, 70, 80, and 90. Its application was faithful to the requirements of the patent.

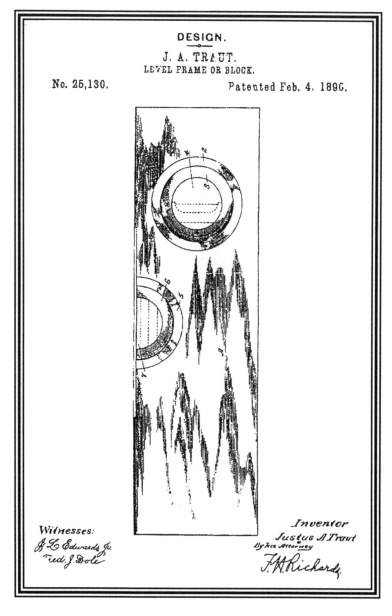

DESIGN FOR A PLUMB AND LEVEL FRAME

For information on Laroy Starrett, see his patent of May 6, 1879 on page 47.

This design patent is for the classic Starrett cast iron level stock. However, the drawings show the central level vial sitting on two short pillars. There is a set screw in the side of each pillar that is used to lock the studs in place. The studs pass through the pillars from the bottom and lock the vial cartridge in place.

Most Starrett cast iron bench levels incorporate most of the elements of this design.

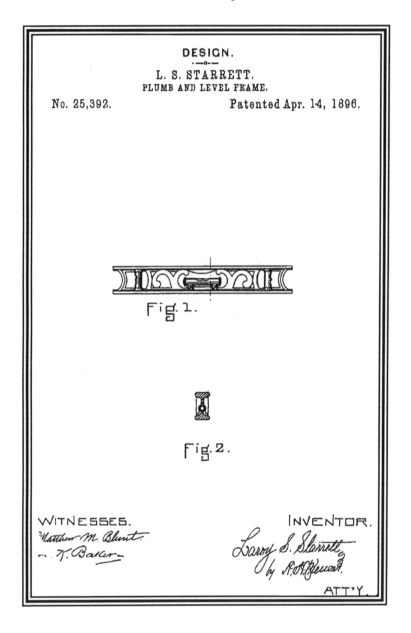

MACHINE FOR GRADUATING GLASSES OR TUBES

Christian Bodmer was co-holder of the patent with Justus Traut. Bodmer had been a Stanley employee since 1887, at least, but was listed as a contractor from 1898 to 1902. In 1903, he was shown as a foreman at Stanley Rule & Level.

For information on Justus A. Traut, see his patent of October 6, 1868 on page 28.

The object of the patent is the creation of a device that (1) automatically places a mark on the high side of a curved vial; (2) always makes it the same length; (3) automatically chooses the proper place to make the mark; and (4) automatically makes the mark to the proper depth. It is interesting to note that the operator still loads the machine by hand, one tube at a time; and that the operator hand rotates the vial to bring the high part of the curve in contact with the cutter. A cast iron cutter is specified.

This was the patent for a machine to mark level glasses. The machine abruptly changed the look of levels. Now, instead of hand marked, inked, or scratched lines, vials became uniform and contained either one or two lines always at the same place.

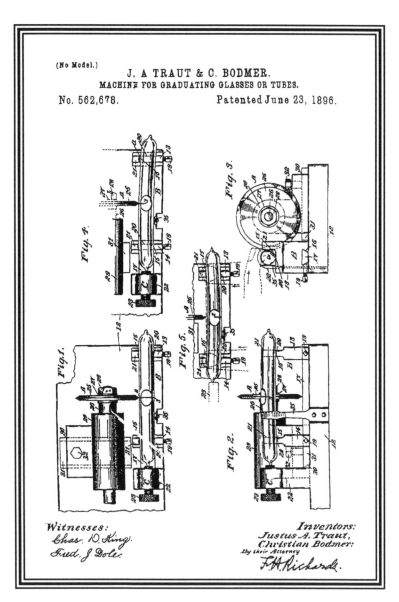

(No Model.)

J. A TRAUT & C. BODMER.
MACHINE FOR GRADUATING GLASSES OR TUBES.

No. 562,678. Patented June 23, 1896.

Witnesses:
Chas. D. King.
Fred. J Dole.

Inventors:
Justus A. Traut,
Christian Bodmer:
By their Attorney
F.H. Richards.

METHOD OF MARKING LEVEL GLASSES OR ANALOGOUS ARTICLES

For information on Justus A. Traut, see his patent of October 6, 1868 on page 28.

For information on Christian Bodmer, see his patent of June 23, 1896 on page 121.

The heart of the technique described here appears to be the use of an extremely thin disk of soft cast iron rotating at very high speed. In addition, the tube to be marked is vibrated back and forth to cause the line to have some breadth. In use, a great deal of heat is generated from friction and this causes the glass to melt and then refuse. However, it refuses containing some black particles from the wheel mixed with the glass. This imparts the colored line. The glass actually gains strength in the area of the mark because of the cooling caused in part by contact with the fluid in the vial. (Other ways of marking by means of a groove substantially weakened the glass, causing it to be very brittle along the line of the groove.)

This is the patent for the technique of lining or marking glass. It was used by Stanley on its spirit vials.

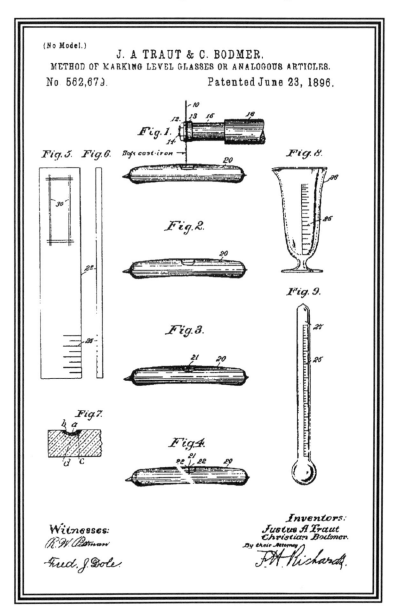

APPARATUS FOR ADJUSTING SPIRIT LEVELS

For information on Justus A. Traut, see his patent of October 6, 1868 on page 28.

The patent describes a portable device for testing and adjusting levels. The requirements for this device are quite simple. The device must possess a rigid flat surface, which will hold its shape. It must have a simple means of adjusting that surface until it is perfectly level. It must retain the quality of being level. This device meets all those requirements. A flat bar is carried on a base stock. It is pivoted on one end, as shown on the left side of *Fig.2,* while the height of the other end is adjustable, as shown by *C.*

This is the cast iron apparatus for use by stores in adjusting levels. Existing examples are quite faithful to the dictates of the patent.

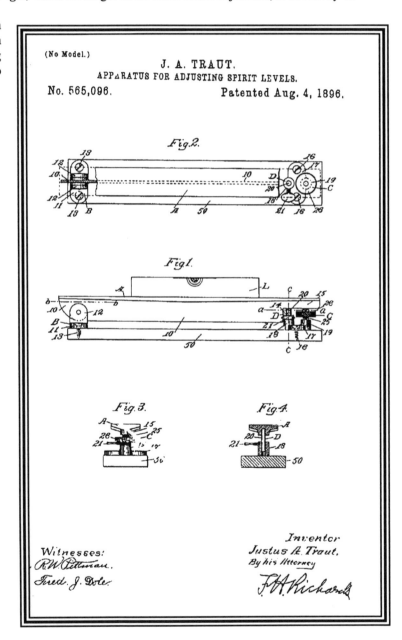

(No Model.)

J. A. TRAUT.
APPARATUS FOR ADJUSTING SPIRIT LEVELS.
No. 565,096.   Patented Aug. 4, 1896.

Fig.2.

Fig.1.

Fig.3.   Fig.4.

Witnesses:
R.W. Pittman.
Fred. J. Dole.

Inventor
Justus A. Traut.
By his Attorney
F.H. Richards

LEVEL

For information about Justus A. Traut, see his patent of October 6, 1868 on page 28.

The patent suggests a level consisting of two parts, as shown in the drawing. (The top part does not appear to have great utility when it is removed from the base.) As made, the top part of the level pivots about a central point in the base (labeled *19* in *Fig.1*) and could be locked in position by using the two screws *21* and *22*. Thus, this level can not only be adjusted to read level, but can be adjusted to carry some slight slope, as shown in *Fig.2*. More importantly, perhaps, is the specification for the rotating vial cover, sometimes called the "improved eclipse cover," that was used on several levels for many years.

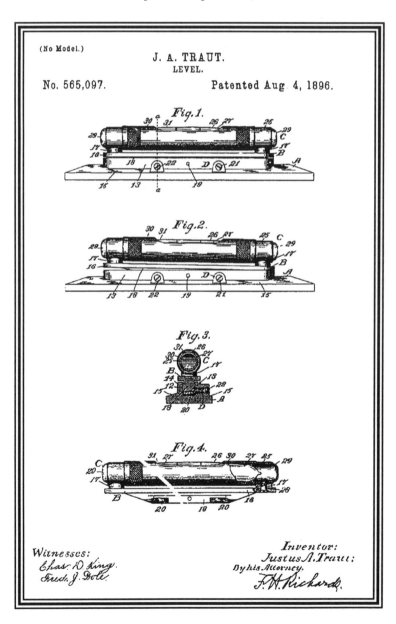

This patent was incorporated into the Stanley № 34 (type 1). The design was changed after being used for only a year, so this level should be considered very scarce.

LEVEL

For information on Justus A. Traut, see his patent of October 6, 1868 on page 28.

The patent describes a slotted stock and a detachable level. The stock could be used in the same fashion as a standard surface level or be attached to a rule or a square. Clamps in the side of the stock would be used to hold it as an attached piece.

The patent also describes a polygonal-shaped level case - preferably a hexagon. Dimples were to be made at the center of the end caps. The level was to be supported in the frame by conical pointed screws resting in the dimples.

The interesting part of the patent is claim № 7, which is a claim for the invention of a hexagonal level with conical end caps. The hexagonal level was not a new device, having been introduced by Richardson (with round or conical end caps) nearly 10 years earlier and used later by Davis. Thus, it is difficult to understand the validity of this claim.

This patent is the basis for the Stanley № 33 and for the short lived № 33½, and these levels are quite faithful to the patent. The support or suspension part of the patent was used on Stanley levels №s. 36, 37 and others. Early Stanley №s. 36 and 37 had this patent date cast into the frame so it is the axial method of support that is claimed for those levels.

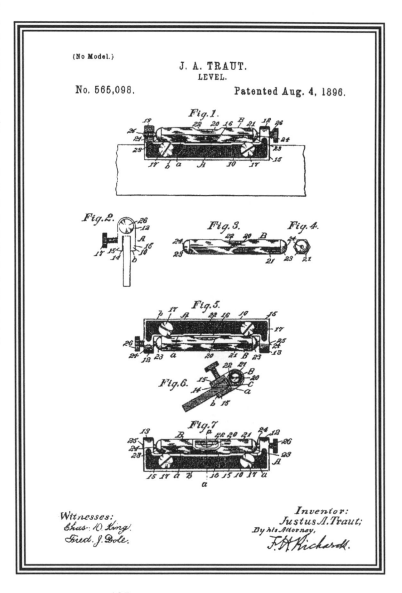

-125-

SPIRIT LEVEL AND PLUMB

According to the New Haven, CT City Directories of 1894-97, William M. Morton operated "William Morton & Co.", manufacturers of special hardware. He moved to Essex, CT in 1900 and nothing further is known about him after this time.

The patent describes a stock containing spirit vials in tubes. There is a central pivot on each tube and a threaded cap on each end. In order to adjust the vial, one cap is turned further onto the tube, while the other is turned off from the tube by an equal amount. The tubes can be locked in place by turning the caps in such fashion that they wedge tight against the frame. Proper vials for this level have a single heavy line in the glass.

This patent was used by "Rollis" in making a level; however, the identity and location of Rollis is unknown.

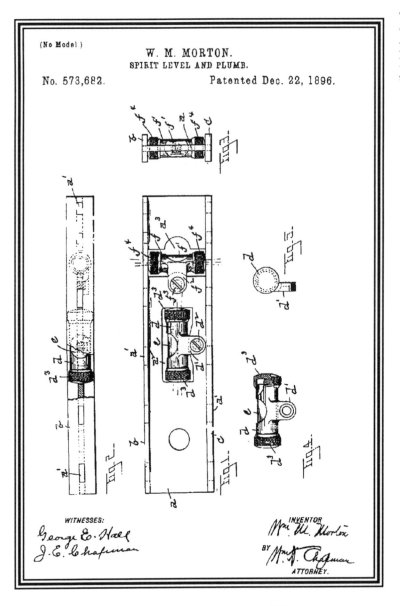

(No Model)

W. M. MORTON.
SPIRIT LEVEL AND PLUMB.

No. 573,682.   Patented Dec. 22, 1896.

WITNESSES:
George E. Hall
J. E. Chapman

INVENTOR
Wm. M. Morton
BY M. J. Chapman
ATTORNEY.

SPIRIT-INCLINOMETER

For information on William M. Morton, see his patent of December 22, 1896 on page 126.

This patent is for a vial fixture that rotates inside a graduated circle. (The frame will be described in Nº 591,139.) In principle, this inclinometer is not greatly different from that of the 1877 patent of Leonard Davis.

The revolving portion, containing the glass level vial, fits closely in a circular hole in the frame. A flat spring is provided to bear against the revolving part, so as to introduce enough friction to keep this portion from moving from the set point. Two adjustable detents are provided, by the use of screws, which bear against the revolving part. The graduations are on the revolving part and the pointer is stationary on the frame of the device. The revolving part is formed of two flanged, circular sections screwed together with the frame between them.

No examples of this inclinometer have been identified to date.

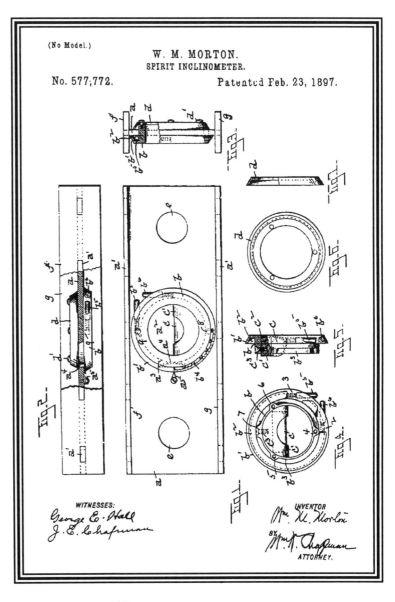

William M. Morton
New Haven, Connecticut

May 4, 1897
581,938

INCLINOMETER

For information on William M. Morton, see his patent of December 22, 1896 on page 126.

Morton envisions a frame consisting of two pieces of sheet metal bent, as shown in the drawing, and riveted together. The level case is attached to a metal piece that is pivoted at the center of the cutout in the frame. The level case can only move through a 90° arc and screws are provided on each end of it for adjusting the travel. The moveable portion is graduated and the pointer is stationary on the frame.

This was Morton's second inclinometer patent; however, the two applications were filed on the same day. It is sheet metal with a simpler vial unit. There is, to date, no evidence that this patent ever resulted in a manufactured product.

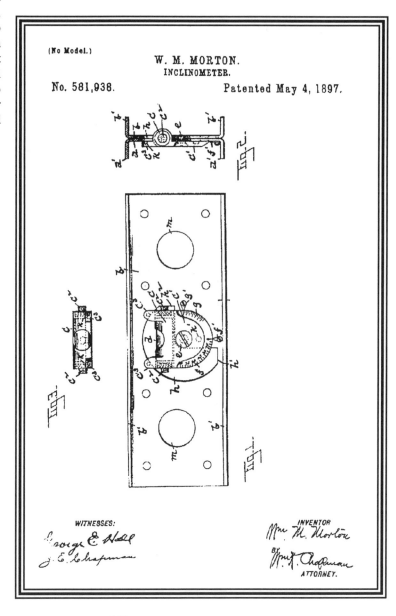

Stephen H. Bellows
Athol, Massachusetts

May 11, 1897
582,517

SPIRIT-LEVEL

For information on Stephen Bellows, see his patent of March 11, 1884 on page 60.

This patent is the Standard Tool Company's answer to the C.F. Richardson patent № 570,056. Said to be useful for the horizontal and vertical adjustment of shafting, the metal level incorporates a V-shaped metal leaf on one end that could be adjusted vertically by loosening a screw and manually moving the adjustable plate. The level itself also incorporates a V-shaped bottom, signaling that it was intended for use on shafting.

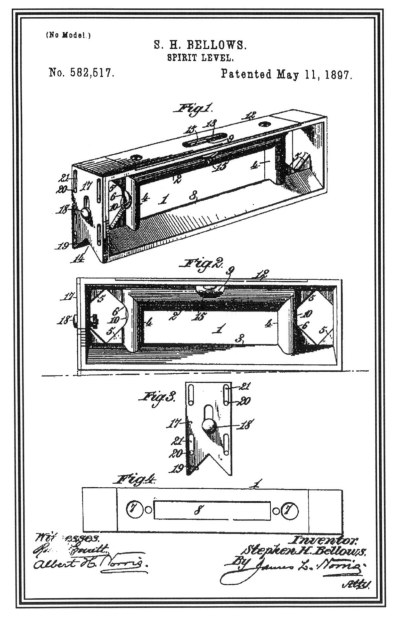

This tool was produced by Standard Tool Company and is well known. It is generally like the patent drawing except that the four guide screws are omitted. The design of the frame is faithful to the Standard Tool Company bench level design.

COMBINED PLUMB AND LEVEL

For information on William M. Morton, see his patent of December 22, 1896 on page 126.

This plumb and level patent covers the type of stock claimed in his first two patents. The claim is for a three piece stock held together with mortise and tenon joints. Both of the first two patents defer to this one for the claim on the stock. However, this patent was granted 10 months later than the first patent and Rollis did not mark its levels with this patent. This implies that the levels made under the December 22, 1896 patent were all made before the granting of this patent. They are relatively scarce levels, which implies that there was only a very short production run of the adjustable level.

No examples of this non-adjustable level have been identified to date.. Metal levels offered at this time by other manufacturers such as the Springfield Level & Tool Company usually included both an adjustable plumb and level and a non-adjustable plumb and level of a somewhat similar design to that detailed in this patent.

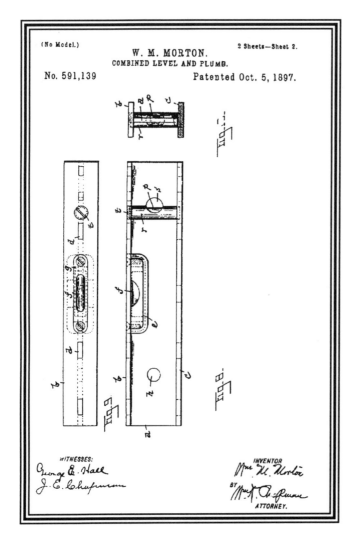

Christian L. Berger
Boston, Massachusetts

October 5, 1897
591,153

ENGINEER'S OR SURVEYOR'S INSTRUMENT

Christian Berger was a German, trained in the construction of scientific instruments, who emigrated to America in 1866. He and George Buff established Buff & Berger in Boston in 1871 and operated it until 1898. At that time he formed C. L. Berger and Sons and operated it until his death in 1922.

This patent is of interest not because of the surveyor's instrument per se, but because of the specification of a two-chambered spirit vial tube (see *Figs.8 - 10*) that is claimed to have superior sensitivity. It is ground on a radius of 300 to 500 feet. Berger recommends the use of a very quick acting fluid such as sulfuric ether (?) or ether. The problem with these materials, in a standard vial, is that they have a very large coefficient of thermal expansion. This causes an improper sized bubble. Thus the second chamber, fused to the main chamber, and of the shape shown, is available to exchange fluid for air and adjust the size of the bubble. Two-chambered vials are not unknown, but previously they resulted from two pieces being cemented together; there the cement is susceptible to attack by the fluid.

This patent also suggests a radial slit in the wall between the chambers rather than the previously used open segment as shown in *Fig.10*.

The author is unaware if this chambered vial was ever produced. It is the vial that is of interest in this patent and not the instrument itself. This interest arises because of a subsequent Traut patent with the same general premise.

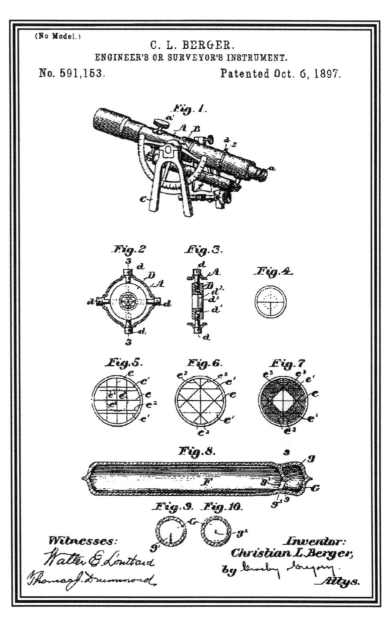

(No Model.)

C. L. BERGER.
ENGINEER'S OR SURVEYOR'S INSTRUMENT.

No. 591,153.   Patented Oct. 6, 1897.

Fig. 1.

Fig.2   Fig.3.   Fig.4.

Fig.5.   Fig.6.   Fig.7.

Fig.8.

Fig.9.   Fig.10.

Witnesses:
Walter E. Lombard
Thomas J. Drummond

Inventor:
Christian L. Berger,
by Crosby Gregory
Attys.

Joseph Carriere
Worcester, Massachusetts

October 26, 1897
592,537

LEVEL

Joseph Carrier was a British citizen at the time of this patent. He was a carpenter living in Worcester and working for W. E. Putnam, a contractor and builder. He was not listed in the Worcester City Directory in 1896, but appears to be there beginning in 1897 through 1908.

This invention envisions a flat bottle shape with the neck of the bottle being inserted into the casing stock, so that the bottom of the bottle faces out. The bottle is sealed with a cork, and the cork is forced all of the way in until it contacts the inside face of the bottom of the bottle. The bottle is to be graduated around the side faces of the bottom such that the graduations can be seen through the bottom of the bottle.

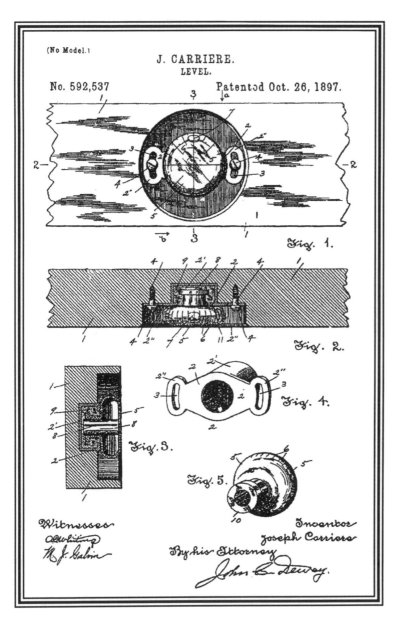

Sufficient air is left in the bottle that when the stock of the level is laid down, such that the bottle is upside down, the device functions as a surface or plane level. When the stock is laid such that the bottom of the bottle is in a horizontal plane, the device functions as a plumb level and could be used as an inclinometer of sorts.

To date, no such device has been identified as having been manufactured.

-132-

LEVEL

In 1898, Charles Mortimer Potter was listed as contractor in the Waterbury-Naugatuck City Directory. In 1901, he was listed as mason and builder. There is little else known about him at this time. A package insert describing the use of the tool shown below was presented in Volume 1 of *American Levels and Their Makers*. The instruction sheet describes its application to fireplace building.

This patent describes an inclined surface, with variable pitch, that attaches to the bottom surface of the level with adjustable fixtures, that extend through the body of the level. The bottom plate of this tool can be lowered by turning screws in the top surface of the level. The bottom plate is attached to the screws through pivot points that allow the two ends of the plate to independently descend different distances. The bottom plate provides a stable base for using the tool in either the plumb or leveling position.

A Potter patent level is known to exist and to be in substantially the same form as shown in this patent. This feature was found on a Stanley level, but it is not known whether or not Stanley was the manufacturer of the total piece.

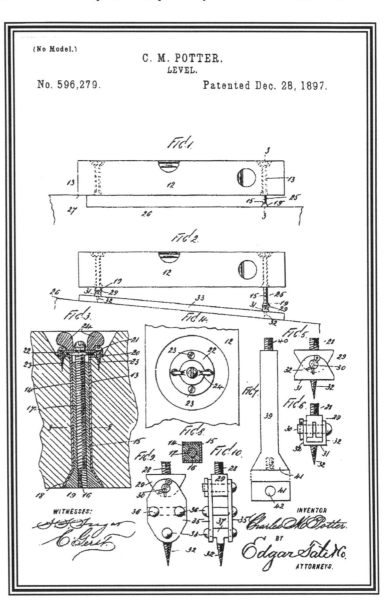

(No Model.)

C. M. POTTER.
LEVEL.

No. 596,279.

Patented Dec. 28, 1897.

COMBINED LEVEL, SQUARE, AND PLUMB

For information on Laroy Starrett, see his patent of May 6, 1879 on page 47.

This is the classic Starrett try-square containing three spirit level vials arranged as shown in the drawing. The tool could be used as a plumb or level as well as a square. This was sold as Starrett's Cross Test Level and Plumb № 134.

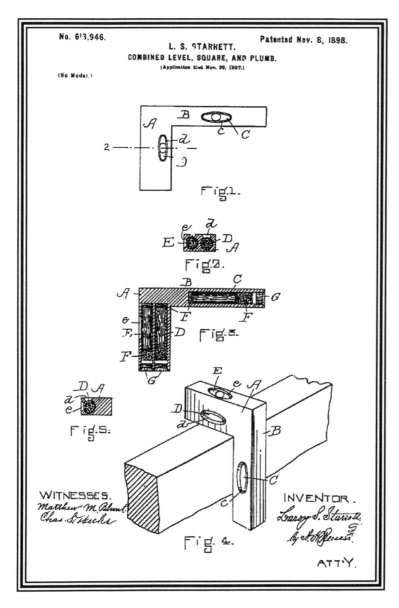

No. 613,946.
Patented Nov. 8, 1898.

L. S. STARRETT.
COMBINED LEVEL, SQUARE, AND PLUMB.
(Application filed Nov. 29, 1897.)

(No Model.)

Fig. 1.

Fig. 2.

Fig. 3.

Fig. 5.

Fig. 4.

WITNESSES.
Matthew M. Blunt
Chas. S. Kuchs

INVENTOR.
Laroy S. Starrett
by A. R. Reuss
ATT'Y.

DESIGN FOR A COMBINATION SQUARE

Burnside Sawyer was born, in 1861, in Templeton, MA. Sawyer began working for Starrett soon after Starrett went into business. It appears that Sawyer then went into business for himself in 1895. He incorporated his company in 1898 and moved to Fitchburg. Sawyer was Treasurer of the new corporation. The company reorganized in 1902, and Sawyer sold his interest at that time.

Like other combination squares, this one incorporates a spirit vial. The main feature of the design is claimed to be a circular center of the web with its raised or beaded edge..

The patent used for Sawyer's Nº 39 Combination Square.

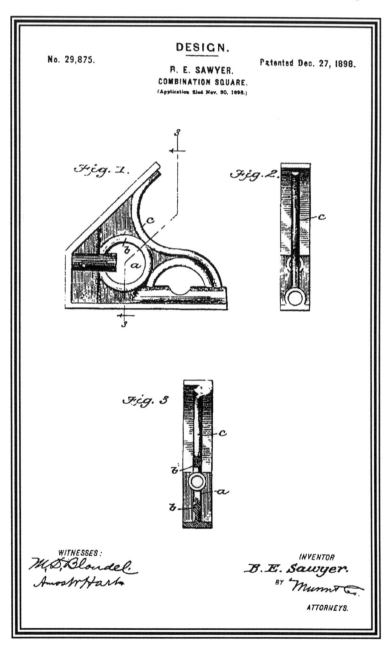

DESIGN.

No. 29,875.

R. E. SAWYER.
COMBINATION SQUARE.
(Application filed Nov. 30, 1898.)

Patented Dec. 27, 1898.

Fig. 1.

Fig. 2.

Fig. 3.

WITNESSES:
M. D. Blondel.
Amos W. Hart.

INVENTOR
B. E. Sawyer.
BY Munn & Co.
ATTORNEYS.

Nathaniel C. Merrill
New York, New York

March 21, 1899
621,358

Assigned to Uncle Sam Novelty Co. of Providence, R.I.

LEVEL

The Uncle Sam Novelty Company was not listed in most of the City Directories of the period. The exception was The Rhode Island Register and Business Directory of 1899. It was only here, in 1899, that The Uncle Sam Puzzle Company was listed. It was located at 47 Washington St., Room 10. Prior and subsequent directories did not reveal any relationship of Merrill to Providence, R.I.

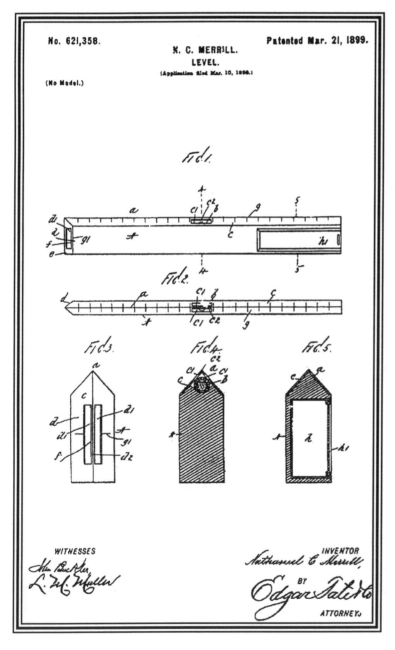

The patent is for a ruler with a triangular ridge on one long edge and one short edge. Spirit level vials are set into the triangular portions of the ruler. A pencil storage compartment is found on one face.

This level patent is appropriately assigned, but no examples of this device are known at this time.

CLINOMETER

According to New Milford City Directories, Frank Cable was a farmer during the period 1888 to 1896, after which, in 1897, he moved to Philadelphia. There were no subsequent listings and Philadelphia City Directories are beyond the scope of this book. The reason for listing New Milford, Connecticut, as his city of residence in 1900 is not understood.

The patent is for a clinometer consisting of a tube or pipe around the inside perimeter of a level stock having both ends terminating in a smaller diameter U tube located in the center of the stock. The large tube is filled about half full with mercury, while the U tube contains alcohol. The space between the two liquids is filled with air. The general idea is that a small movement of mercury in the large tube will cause a large movement of alcohol in the small tube.

Any desired type of scale can be viewed behind the U tube. The gradations on the scale would depend on the ratio of diameters of the two tubes. It is not at all clear from the text of the patent what is to prevent the mixing of the liquids during an upset.

Cable envisioned his device to be a ship's inclinometer. To date, no examples of a device resembling this patent are known.

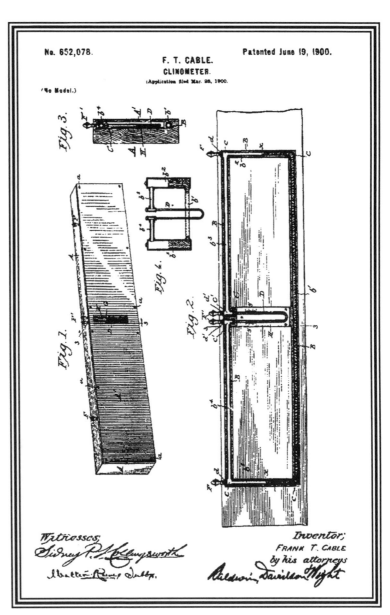

SPIRIT-LEVEL AND PLUMB

George F. Quinby was a principal in Atlantic Machine Screw Company of Boston, MA.

The main concept of the patent is to suspend the level case from a pivot point on each end such that the level stock can be placed in any position (including upside down) and the top of the level case will be upright. Both a plumb and level can be thus equipped. The stock shown in *Fig.4* features the Starrett design, although there is no evidence that Starrett collaborated with Quinby.

The idea for suspension of a level cartridge seems practical, but no examples have been observed.

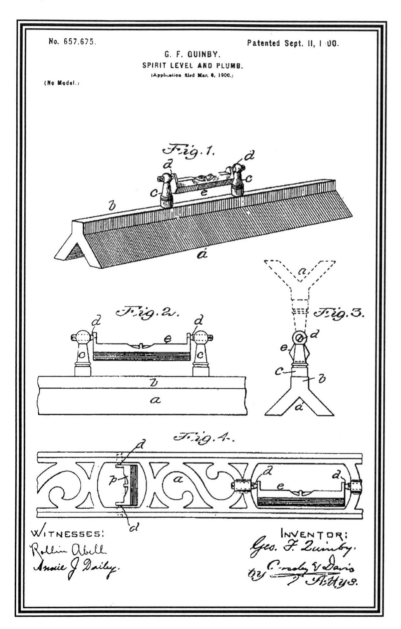

No. 657,675.

Patented Sept. 11, 1900.

G. F. QUINBY.
SPIRIT LEVEL AND PLUMB.
(Application filed Mar. 6, 1900.)

(No Model.)

Fig. 1.

Fig. 2.

Fig. 3.

Fig. 4.

WITNESSES:
Rollin Abell
Annie J. Dailey.

INVENTOR:
Geo. F. Quinby.
by Crosby & Davis
Attys.

John S. Bogardus
Stamford, Connecticut

December 4, 1900
663,252

LEVELING, PLUMBING, AND ANGLE-MEASURING INSTRUMENT

John Bogardus worked as an architect through the 1880s and 1890s, according to the Stamford City Directories. He was last found in the 1902 Directory when he had his office in his home.

The patent claims that this device is to be a leveling, plumbing and angle measuring instrument. The device has a flat base with metal arms arcing up overhead and meeting a head-piece from which is suspended a dial plate. The length of the metal arms is adjustable. On the rear of the dial plate is a sighting tube. A plummet is suspended behind the dial plate and has an indicator pointing to the dial and a compass inside. The circular dial has vertical slits cut into it to be used when plumbing the wall of a building. The two adjustable arms can be rotated to match the pitch of a roof or other slope. The circular dial affair has the sighting tube attached to the back and can be rotated relative to the plummet in order to sight elevations.

Bogardus makes claims that this instrument does anything that a more expensive transit-type instrument could do.

There is no known example of the device envisioned by this patent.

PATENT DRAWINGS CONTINUED ON NEXT PAGE.

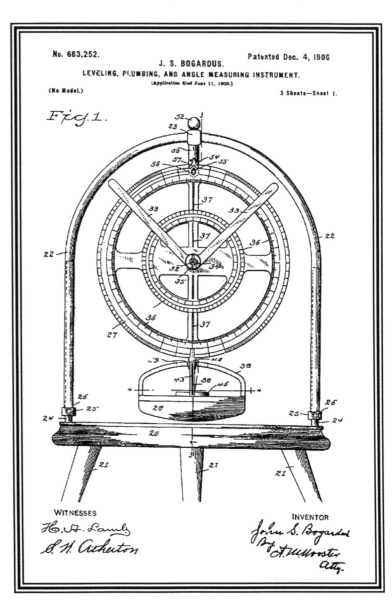

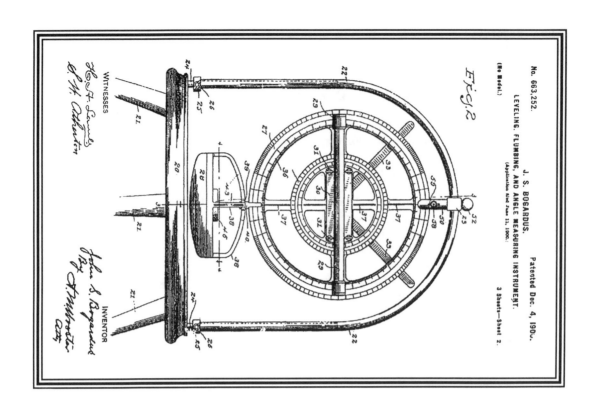

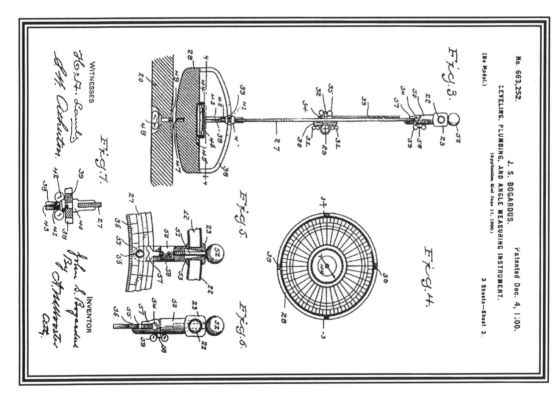

LEVEL-GLASS

For information on Justus A. Traut, see his patent of October 6, 1868 on page 28.

This patent claims that coating the ends of the outside of the tube will cause the edges of the bubble to reflect that color, thus making the position of the bubble more apparent to the viewer.

It is unclear whether or not this patent was ever utilized by Stanley or anyone else, although the author recalls seeing vials with silvered ends.

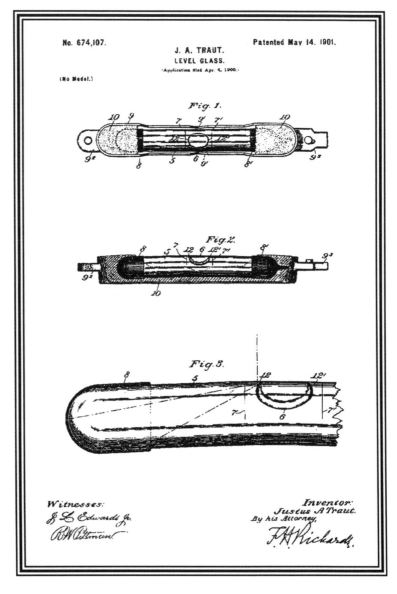

COMBINATION-TOOL FOR SQUARING PURPOSES

Andre Delin was variously listed in the Everett City Directories, between 1897 and 1903, as a bricklayer and a mason.

The patent drawing provides a clear explanation of the patent principles. The stock is an ordinary carpenter's level, with the added provision of a slot for storage of the detachable tongue of the square, and appropriate receptacles to receive the flanges on the tongue.

To date, no examples of this patent have been identified.

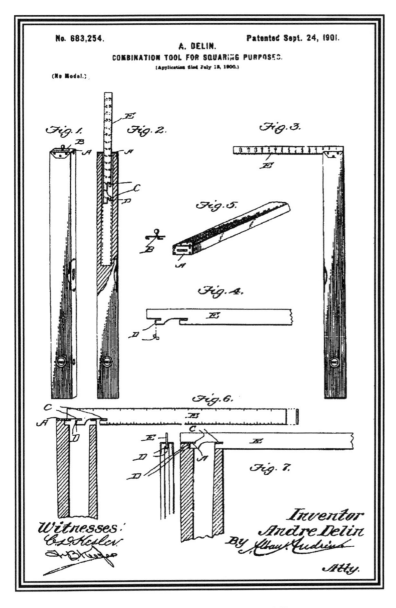

James Bullard
Springfield, Massachusetts

October 29, 1901
685,569

Assigned to Overman Automobile Co. of Chicopee, Mass

GRADOMETER

James Bullard was Treasurer and Manager of the Aerated Fuel Company, located in his home at 777 State Street.

Albert H. Overman was President of Overman Automobile Co., an automobile manufacturer. (In 1900, Overman's company was Overman Wheel Co, a manufacturer of bicycles and automobiles in Chicopee, Mass.) In 1902 Overman was listed as removed to London, but the company business listing was retained and Bullard was not listed.

The claim is that this device will best determine the inclination of a self-propelled vehicle in motion. Use of an air bubble in a tube is said to be subject to too much vibration to be accurately read.

The device comprises a curved glass tube mounted so the ends are higher than the center. The tube is filled with liquid and liquids of various viscosities may be chosen, depending on the need for damping. A metal ball is sealed in the tube and the diameter may be chosen depending on the need for damping. Higher viscosity liquids and larger diameter balls lead to greater damping. The patentee suggests using a steel ball with alcohol for the liquid. The claim is that the ball will seek the lowest point and the movement of the ball will be damped by the liquid. The tube is graduated in percent inclination.

No example of this patent has been identified. However, because of the various industries in which the device may have been used, it might well exist and be unknown to the author.

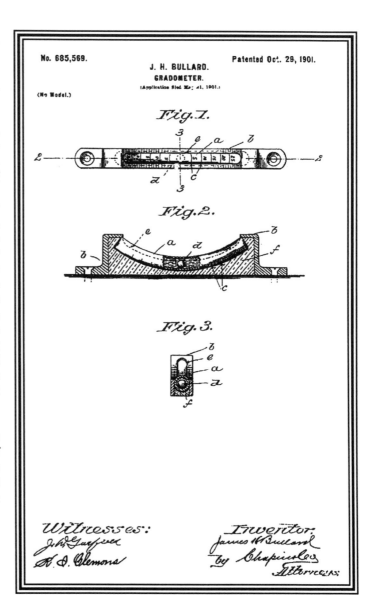

-143-

Harold Kelly
Biddeford, Maine

SPINDLE-PLUMB

Harold and his father John P. Kelly were in the business of making springs (clearer springs) for the weaving mills. John was listed as such until 1896. From 1896 until 1904 (and perhaps after), Harold was simply listed as a machinist.

This patent is for a device to use in determining and adjusting the plumb on spinning spindles while they are operating. Basically, it consists of a portable bobbin that sits on the spindle to be plumbed. The top of the bobbin fits into a box containing a race equipped with ball bearings. An unspecified means is provided to prevent the box (and thus the levels) from rotating. Inside of the box sit what are best described as cross test levels.

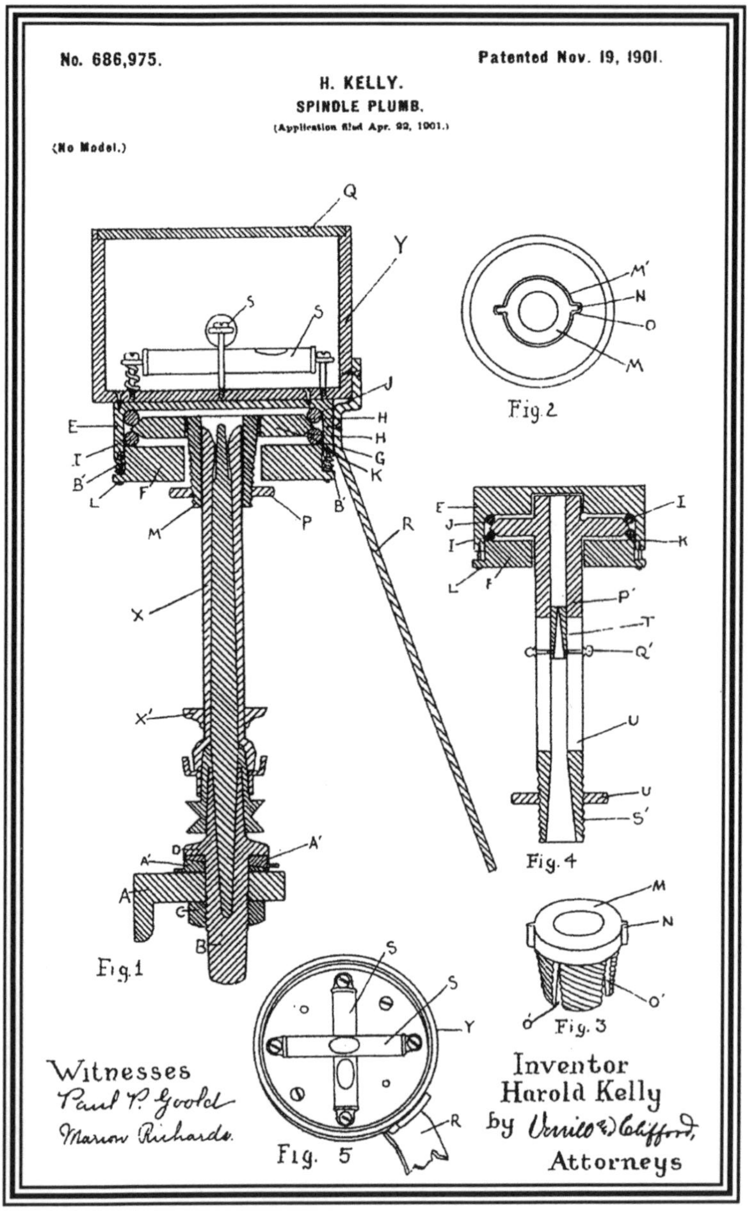

The patent contains directions for changing the plumb of spindle.

The author is unfamiliar with the spinning and weaving trades and is unaware of whether or not this tool was ever manufactured.

No. 686,975.
Patented Nov. 19, 1901.

H. KELLY.
SPINDLE PLUMB.
(Application filed Apr. 22, 1901.)

(No Model.)

Fig. 1
Fig. 2
Fig. 3
Fig. 4
Fig. 5

Witnesses
Paul P. Goold
Marion Richards.

Inventor
Harold Kelly
by Unilo & Clifford,
Attorneys

PLUMB

For information about Carsten Diedrich Janssen, see his patent of November 17, 1885 on page 73.

This patent is for a simple tool, wherein a plumb bob hangs inside a stock. The line holding the plumb bob is viewed through a window and against a scale.

Known examples of this level incorporate a vial cartridge within the stock and viewable through the window. The plumb bob itself shows in the window of the known examples (instead of just the line). The known examples also contain a small take-up reel at the top of the stock that was not specified in the patent.

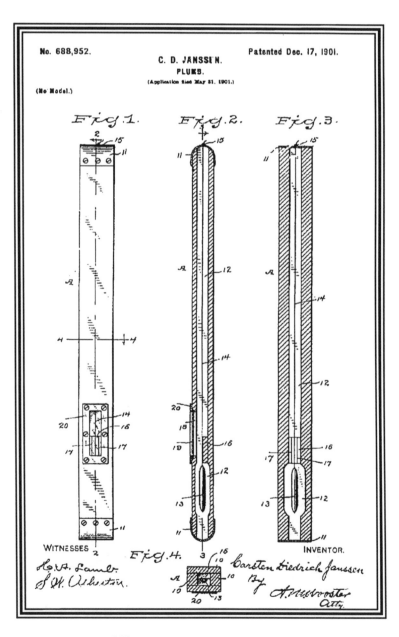

SPIRIT LEVEL

For information on Justus A. Traut, see his patent of October 6, 1868 on page 28.

The patent is for a method of constructing a kind of universal vial assembly that is easily adjustable. Perhaps the most novel thing about this level is the inclusion of the vial guard to protect the vial from damage from the anticipated use of a plumb bob with this level. The slot down the flat face of the level was intended to accommodate the line on the plumb bob.

Figs. 6-8 show an alternative way of fastening the vial carrier.

To date, there are no known levels containing the structures detailed in this patent. It may be noted that the patent descriptions of Justus Traut had been getting more difficult to read toward the later part of his career. This patent and the one granted to him on January 27, 1903 are pretty much the epitome of trivial and wordy description.

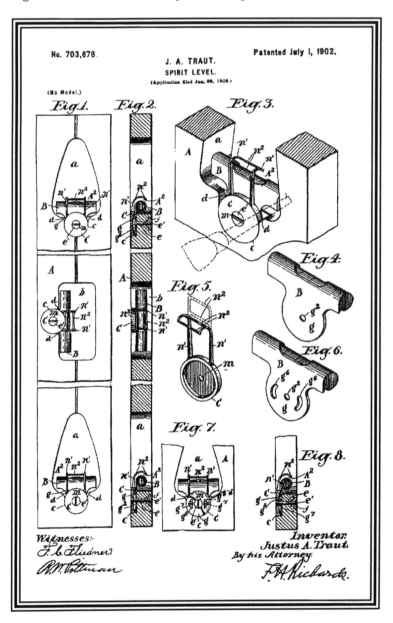

No. 703,678.

J. A. TRAUT.
SPIRIT LEVEL.
(Application filed Jan. 28, 1908.)

Patented July 1, 1902.

(No Model.)

Fig.1. Fig.2. Fig.3. Fig.4. Fig.5. Fig.6. Fig.7. Fig.8.

Witnesses:- F. L. Fliedner, R. W. Pittman

Inventor: Justus A. Traut. By his Attorney, F. H. Richards.

Thomas Seymour Tilley
Newport, Rhode Island

October 21, 1902
711,801

Assigned one-half to William J. Thomas of Newport, RI

MEASURING INSTRUMENT

Thomas S. Tilley was a brass finisher working for Pascal H. Stedman at 111 John Street in Newport, Rhode Island, according to the 1902 Newport City Directory. From 1906 to 1915, Tilley was listed as an electrician working for Scannevin and Potter.

William J. Thomas was an electroplater working at the same location.

The device is an inclinometer with grade level capabilities. The inclinometer is a variety of the pendulum or weighted needle type. The needle or indicator has dual pointers so that angles can be read on scales on either side or through a view port on the top. A channel iron base is hinged and provided with an articulated leg that can be locked into position at any angle. The base can be locked flat to the stock for use as an ordinary level or inclinometer.

To date, no example of this device has been identified.

PATENT DRAWINGS ON NEXT PAGE.

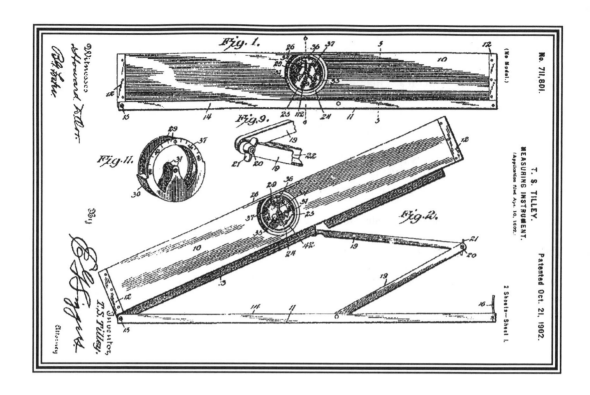

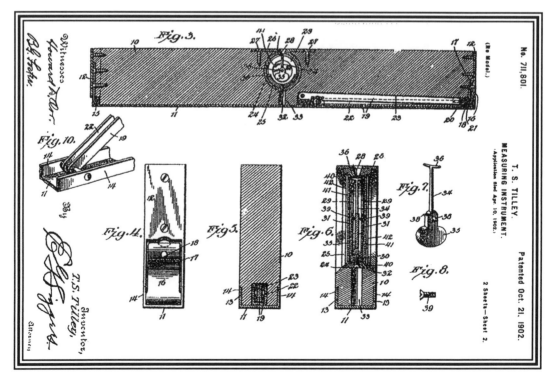

LEVELING INSTRUMENT

For information on Justus A. Traut, see his patent of October 6, 1868 on page 28.

This patent is for a grade level, wherein the plumb and level are locked together and move in tandem when the adjustment is made for grade. The fulcrum of the lever arm is the screw s and the arc of the arm is constrained in the slot $g,$. The scale D is also curved according a circle with its center at s. Although the level vial will move a greater distance than the plumb vial, they will move through the same angle.

There are no known examples of this tool to date. The tool resembles the grade levels produced by L.S. Starrett.

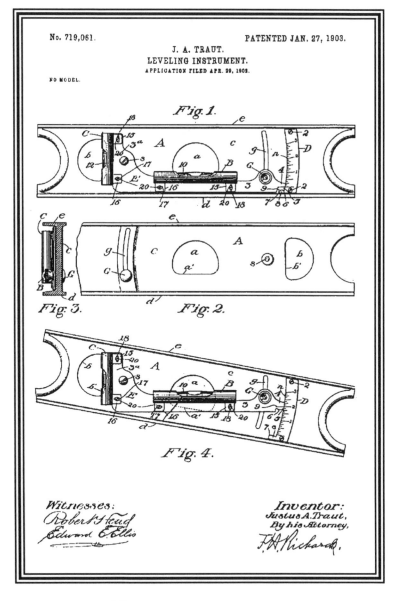

LEVEL

For information on Justus A. Traut, see his patent of October 6, 1868 on page 28.

This patent is for a two-chambered vial to allow changing the size of the bubble. The patent contains 22 claims and each seems to be little more than another way to describe what is obvious after reading the first claim.

To date, no carpenters' levels are known to have vials with the patented kind of chambering mechanism, however, similar mechanisms have been used in various surveying instruments; there the purpose is to effect a kind of temperature compensation for the size of the bubble. A more sophisticated version of this vial was patented in 1897 by Christian L. Berger.

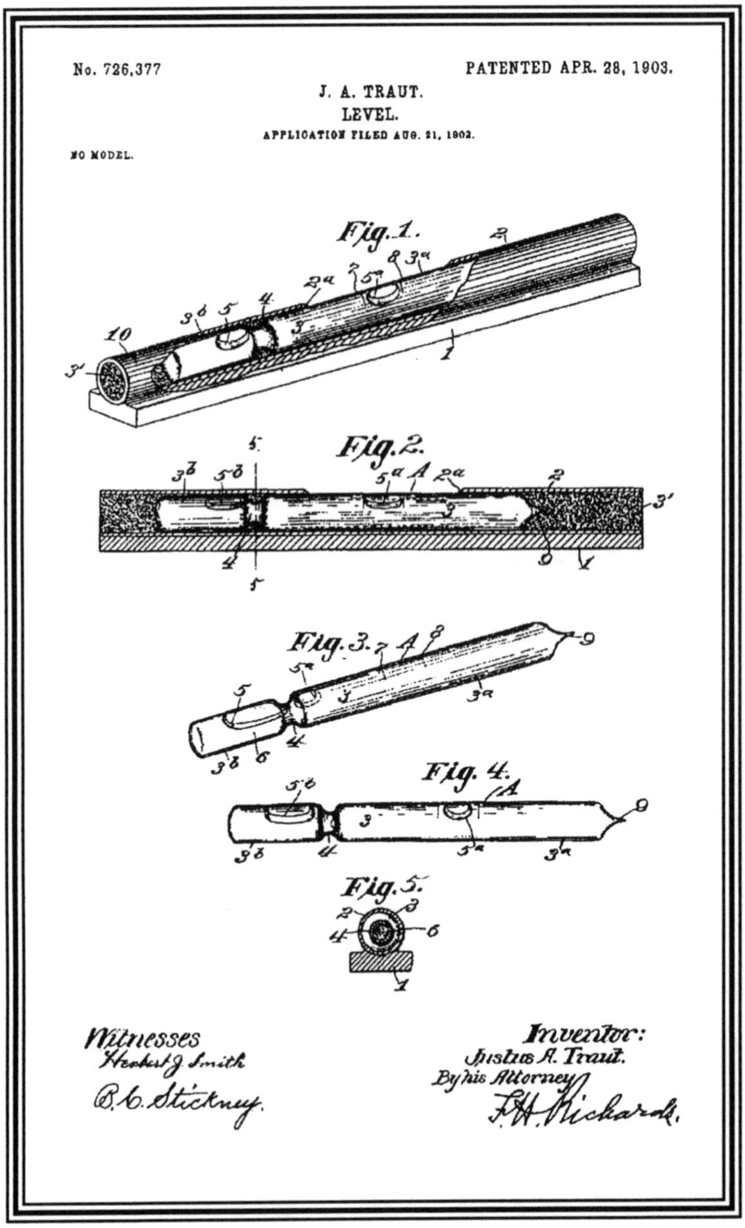

No. 726,377

PATENTED APR. 28, 1903.

J. A. TRAUT.
LEVEL.
APPLICATION FILED AUG. 21, 1902.

NO MODEL.

Fig. 1.

Fig. 2.

Fig. 3.

Fig. 4.

Fig. 5.

Witnesses
Herbert J. Smith
B. C. Stickney.

Inventor:
Justus A. Traut.
By his Attorney
F. H. Richards.

Alonzo E. Rhoades
Hopedale, Massachusetts

June 30, 1903
732,233

Assigned to Draper Company of Hopedale, Massachusetts (A Corporation of Maine)

A DEVICE FOR PLUMBING SPINNING-SPINDLES

Alonzo Rhoades worked for the Draper Company, which made looms and shuttles, and at one time employed about 4000 people in Hopedale, Mass. Rhoades was evidently a supervisor of some type in the machine shop operations. He held other patents for devices such as bobbin holders and spoolers.

Elihu Dutcher was associated with the early Draper Corporation, in that he and his brother invented and patented the original Dutcher Temple. The Drapers purchased Elihu's half interest in the temple in 1854. Elihu used those proceeds to buy a farm in Wisconsin but died of cholera on the second day after his arrival in Wisconsin. Elihu Dutcher is of interest because of his early patent of a metallic plow plane.

This device was meant to operate while the machine was running. The claim is that only the lightest pressure is necessary for sufficient contact to obtain a reading. The blades b^5 attached at the top and bottom of the leveling device are contacts for the spindle. These blades can be adjusted for different lengths from the support b. This is necessary because of the potentially different tapers of the spindles. A handle is attached to an axle through the horizontal center of the device. A spirit level vial is placed in a tube, perpendicular to the main stock of the device, in parallel with the extended handle. The level is read when the two blades each contact the spindle. It may be also be used while the machine is stationary.

The industry that could have employed this level is foreign to most tool collectors and information regarding its production and use is unavailable.

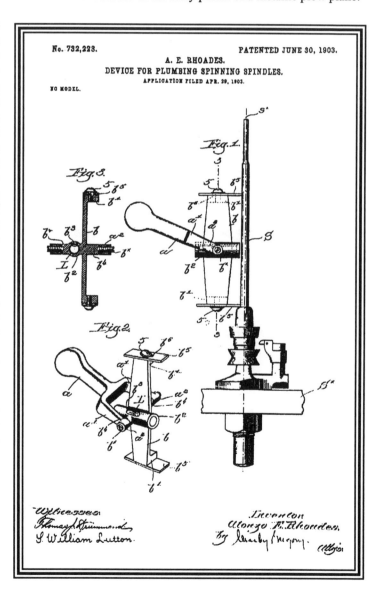

Herbert T. Dillon
Henderson, Maine

September 29, 1903
740,255

LEVEL

Herbert Dillon was born in Spring Hill, New Brunswick, Canada, on October 10, 1876. He emigrated to the U.S. in 1895 and was naturalized in 1919. Various census data showed him to be a proprietor of a barber shop and pool room. He drowned on November 9, 1936. Death records list him as a retired theater operator. Henderson, Maine, is now called Brownville and is in Piscataquis County.

This patent describes the provision for a metal arm or base under a wooden carpenters' level. The arm is hinged at one end of the level and can be moved to provide a slanting base to the level. At the other end of the arm is a finger that slides through a metal channel added to the side of the level. The channel is graduated along its length and the finger can be fixed in position with a knurled nut passing through the finger and the stock.

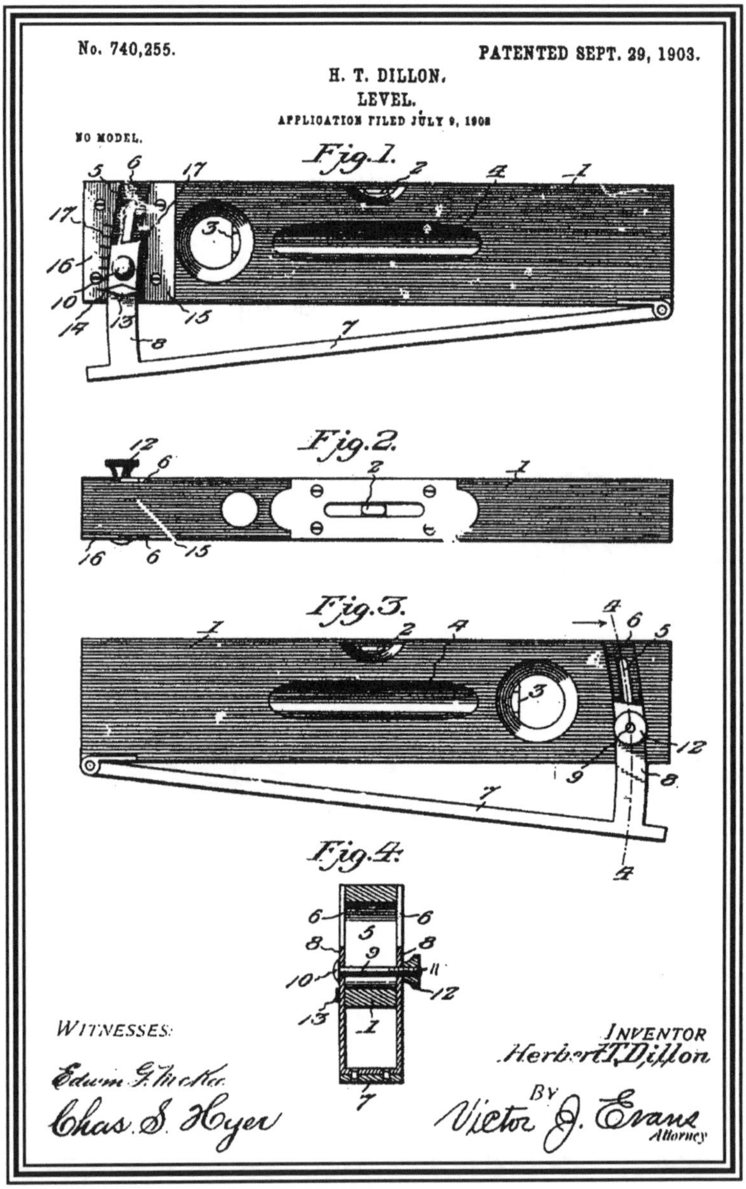

This device could be added to any existing wooden carpenters' level. The use of non-corrosive metal or alloy is specified. (This really means any metal that resists corrosion.)

No examples of this patent have been found to date.

No. 740,255.   PATENTED SEPT. 29, 1903.
H. T. DILLON.
LEVEL.
APPLICATION FILED JULY 9, 1903.
NO MODEL.

Fig.1.
Fig.2.
Fig.3.
Fig.4.

WITNESSES:
Edwin F. Tucker
Chas. S. Hyer

INVENTOR
Herbert T. Dillon
BY
Victor J. Evans
Attorney

Thomas Seymour Tilley
Newport, Rhode Island

October 6, 1903
740,491

Assigned one-half to James Oscar Peckham of Middletown, RI

COMBINED LEVEL AND INCLINOMETER

For information about Thomas Tilley, see his patent of October 21, 1902 on page 147.

No information about James Peckham is available at this time.

Tilley envisions an 18" metal level that is hinged in the middle, so that it opens to form a right angle. The outer edges are ruled. Using a clamp that can engage a curved leg, the two portions of the level can be locked at any angle up to 90°. Semicircular recesses are formed in each leg to receive an inclinometer feature that had been previously patented. [Tilley Nº 711,801, October 21, 1902.] Since the inclinometer is meant to be read through the top of the frame as well as from the sides, the inclinometer portion of this device must be attached to the uppermost of the two legs.

No example of this device has been identified at this time.

PATENT DRAWINGS ON NEXT PAGE.

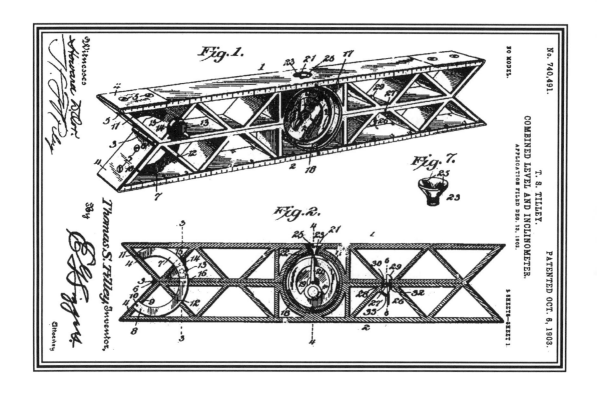

T. S. TILLEY.
COMBINED LEVEL AND INCLINOMETER.
APPLICATION FILED DEC. 12, 1902.

PATENTED OCT. 6, 1903.

2 SHEETS—SHEET 1.

Fig.1.

Fig.7.

Fig.2.

Witnesses

Thomas S. Tilley, Inventor,

By

Attorney

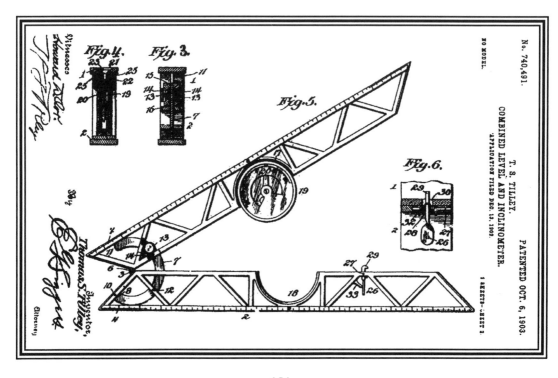

T. S. TILLEY.
COMBINED LEVEL AND INCLINOMETER.
APPLICATION FILED DEC. 12, 1902.

PATENTED OCT. 6, 1903.

2 SHEETS—SHEET 2.

Fig.4.

Fig.3.

Fig.5.

Fig.6.

Witnesses

Thomas S. Tilley, Inventor,

By

Attorney

RULE

James Bush was a steam fitter born in 1864 in New York.

The Bush patent comprises a two-foot rule hinged in the middle, and containing both plumb and level spirit vials in one leg. There seems to be nothing else remarkable about the device.

To date, no example of this patent has been identified.

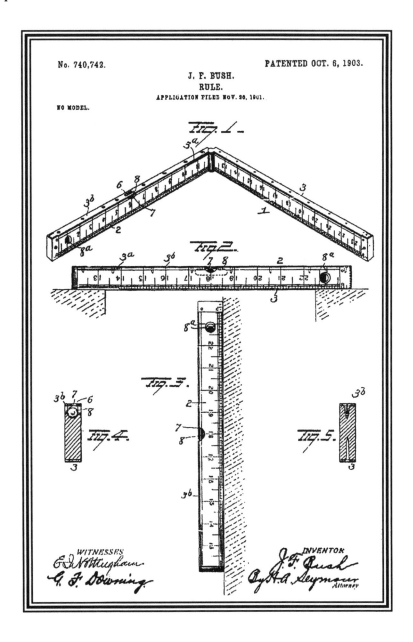

Justus A. Traut
New Britain, Connecticut

May 24, 1904
760,587

Assigned to the Stanley Rule and Level Company of New Britain, CT

LEVEL

For information on Justus A. Traut, see his patent of October 6, 1868 on page 28.

In this device Traut describes a level fixture to be placed in the stock of a wooden level. It is prescribed that the flanges from both sides of the level are directly connected by screws. The assembled device resembles a brass spool with a level vial placed through opposed openings in the spool core. It is intended that adjustments be made by loosening the screws, and using a tool such as a screwdriver, rotating both flanges at the same time.

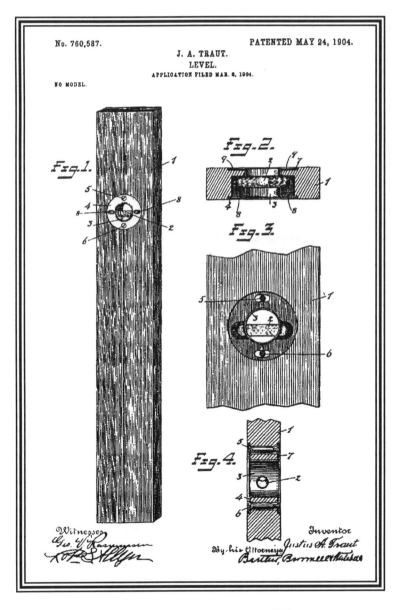

This patent is for an easily adjustable circular brass porthole assembly. The fitting is similar to that found on Stanley № 30 and № 50 levels that employ Traut's March 22, 1890 patent, but these levels have a flange beneath the cover plate on both sides of the stock. The level construction details for the level described in this patent seem similar to levels that have been observed, however, no such Stanley levels are known.

Anson K. Cross
Ashland, Massachusetts

May 2 , 1904
761,033

ARTIST'S LEVEL

No information about Anson Cross is available.

The main piece described by this patent, is a weighted rule as shown in *Fig.1*. The weight hangs below the cross piece and provision is made for attaching it to the rule by means of a pin placed through holes in the rule and the weight. The projections on the spear point on top are meant to be held between thumb and forefinger, and to function as a small axle so that the entire device will function as a plumb bob. The user may gain size perspective using the side scales shown in *Fig.2,* and may add detachable wire squares (as shown in *Fig.3*) into a groove in the large frame, as desired, in order to frame the individual elements.

Fig.4 shows yet another claimed version of the device. In this version, the top cross piece is held to the main rule with a spring, and vertical pieces may be suspended from this cross piece to act as secondary squares. The bottom cross piece is attached directly to the weight and contains bevels to enable the user to frame other components of his picture.

There is no information regarding the potential manufacture of a device according to this patent. The device is unorthodox and required almost two years from submission to approval. While not meeting most of the criteria for a level, it is a form of plumb level and included here for its novelty value.

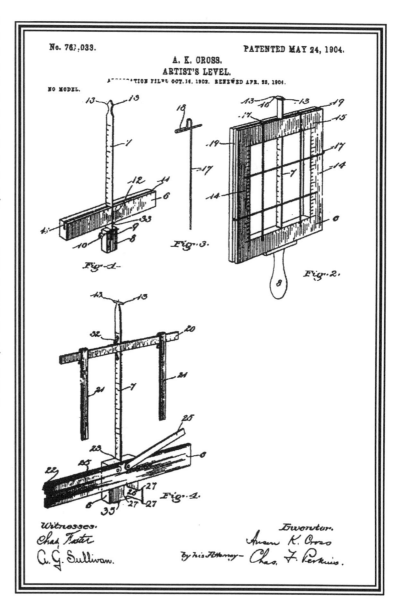

-157-

Peter Lord
Worcester, Massachusetts

June 7, 1904
762,072

LEVELING INSTRUMENT

Peter Lord Jr. had another invention known as the "Lord" Patent Saw File, which he manufactured as a principal in the firm of Lord, Whitter and Park. That was a rather short-lived venture that operated around 1900. In 1895 Peter Lord Jr. was listed in the directories as a carpenter boarding at 2 High Street, which was apparently his father's home. He died in 1907 at the age of 38.

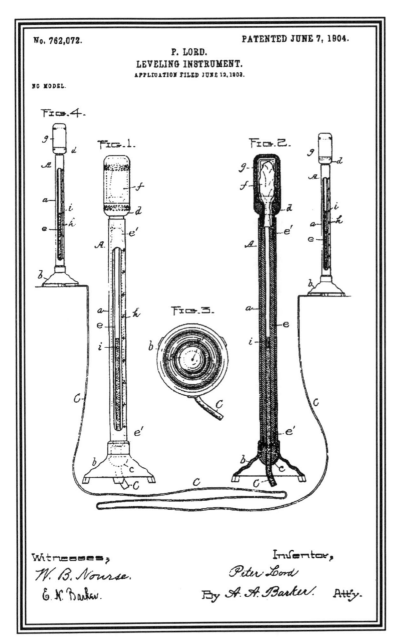

This patent is for a hydrostatic leveling instrument. The stock is a hollow metal tube, slotted for much of its length, and graduated throughout the slotted portion. The metal tube is lined with a glass tube that, on the bottom, passes through a bushing and is connected to a rubber tube. On the top a thin rubber bag (a balloon) is attached to the top of the glass tubing and is contained, for protection, in a metal can-like device. There is a hole in the top of the can to allow for movement of air as the balloon expands and contracts..

The balloons serve a dual purpose. They prevent the loss of fluid at the top of the tube, and allow for the expansion of air from the glass tube.

Lord points out that the connection between the two end tubes could be a pipe of any geometry, and the same result would be achieved.

To date, no examples of this device have been identified. In reality, the level stocks should be relatively easy to produce.

COMBINED SQUARE AND BEVEL PROTRACTOR

For information on Laroy Starrett, see his patent of May 6, 1879 on page

Like his first patent, this device utilizes a metal rule and, with it, a metal head. This head, however, has two flat surfaces (arms) at 90° to each other. Each arm contains a spirit level vial and is slotted from end to end to form a blade seat. There is a graduated rotating part to the head that contains the rule allowing the device to serve as a protractor. As with his other device, the head contains a hook bolt to engage and retain the rule.

This patent covers Starrett's "Patent Protractor № 16".

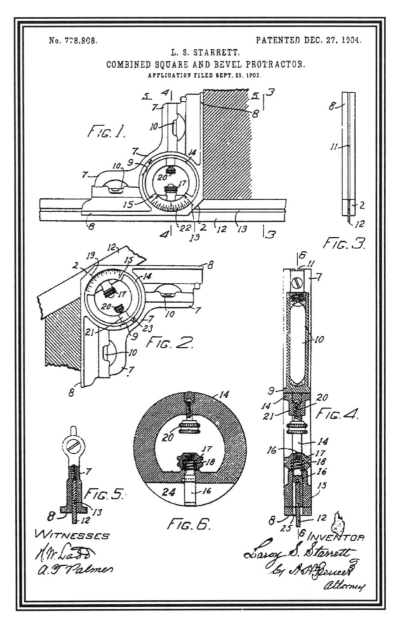

No. 778,808.

PATENTED DEC. 27, 1904.

L. S. STARRETT.
COMBINED SQUARE AND BEVEL PROTRACTOR.
APPLICATION FILED SEPT. 25, 1903.

FIG. 1.

FIG. 2.

FIG. 3.

FIG. 4.

FIG. 5.

FIG. 6.

WITNESSES

INVENTOR
Laroy S. Starrett
by
Attorney

INSTRUMENT FOR SQUARING STONE &c.

Fred T. Stevens, an engineer, died in October 1935. The census places him in Woodstocktown, Maine.

This device comprises a 45° segment of a circle, sitting on a base. The segment is graduated along its outer edge, and has a slot parallel to the outer edge. At the center of the circle, from which the segment was taken, there is an arm of the same length as the base. At the end of the arm are two spirit vials at right angles to one another. The arm can be fixed in position by a screw passing through the slot in the quadrant.

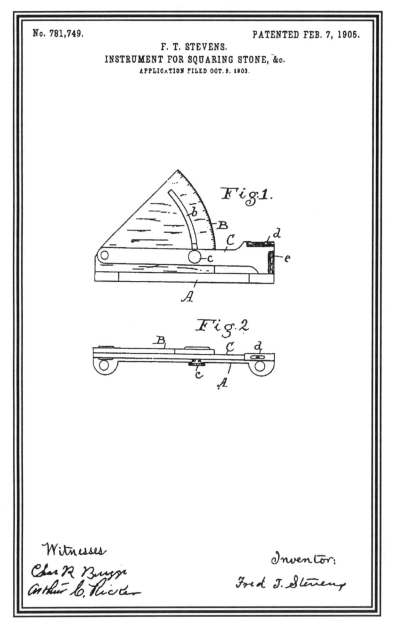

No. 781,749.   PATENTED FEB. 7, 1905.
F. T. STEVENS.
INSTRUMENT FOR SQUARING STONE, &c.
APPLICATION FILED OCT. 9, 1903.

Fig.1.

Fig.2

Witnesses

Inventor:
Fred J. Stevens

In use, the base is placed on one side of a stone and the inclination noted. The opposite face can then be dressed to the same inclination, thus giving the stone parallel faces.

In the only example of this tool observed to date, the arm movement appears restricted to 30°. The tool is well marked with the patentee and patent date cast into it.

ILLUMINATED SPIRIT-LEVEL

Raymond O. Stetson was the son-in-law of Edwin Stratton. Stetson purchased the Stratton Bros. Level Co. from his father-in-law on February 15, 1902. Stetson sold the Stratton Level Co. to the Goodell-Pratt Co. in May 1912. He continued to be active in the tool making business at Goodell-Pratt. From 1915 to 1929, he was a foreman at its factory. Stetson apparently left Goodell-Pratt in 1930, because after that time the Greenfield City Directories listed him as a cabinet maker.

This device is, as the name implies, an illuminated level comprising two bulbs and a battery. Most of the text of the patent is used to describe the circuit and the switches and the means of replacing the batteries. As shown in the patent drawing, the battery chamber is on the end of the stock, the bulb to light the plumb vial is placed opposite the vial in the porthole, and the bulb to light the level vial is placed beneath the vial. The switch is in the center of the side of the stock. Current is conducted through the surface plates *25* meaning that it is not necessary to bore holes through the entire length of the level stock. The switch may be a push button type, in which case both vials are lit at the same time. When the switching is done with the lever as shown in *Fig.1,* the vials may be lit individually.

This level was manufactured, and it is advertised in the Stratton Bros. Catalog as № 101 or 110 depending on the amount of brass trim. The level, except for the battery and lights, followed a pretty standard Stratton Bros. format. It was one of the earliest lighted level patents to be obtained, and one of the very few to be manufactured. This patent required 2½ years to be approved.

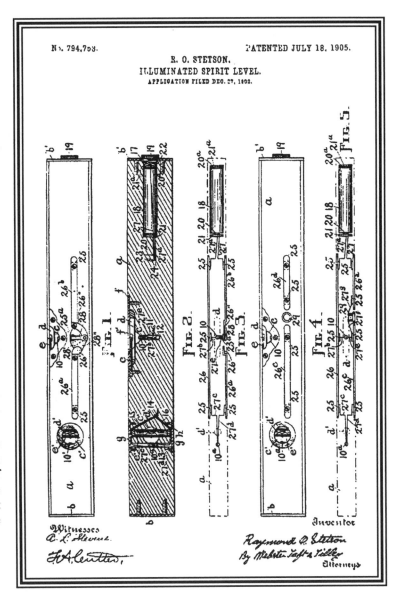

Albert F. McDowell
Malden, Massachusetts

December 12, 1905
806,987

LEVEL

Albert Frederick McDowell was a brick mason who, it was said, had several other inventions relative to his trade. This was apparently the most notable. He immigrated from Canada as a young man and died in Malden at age 78.

The addition of McDowell's patent device turns a carpenter's level into a grading level. A second porthole is added, and into this is inserted a hollow cylinder containing a spirit vial fixture. The cylinder is attached to a long indicating arm with a pointer on the end. The pointer references against a short scale let into the side of the level. A set screw on the end of the pointer rides in a channel near the scale and can be tightened to maintain a desired setting. The position of the scale is adjustable for calibration purposes.

This level was produced by Stratton Bros. and Stratton Level Company in several configurations, ranging from the fine brass bound Stratton configuration to an unbound version with little brass and a poorer quality mahogany. In the level as manufactured, the indicating arm is much longer than implied in the patent drawings.

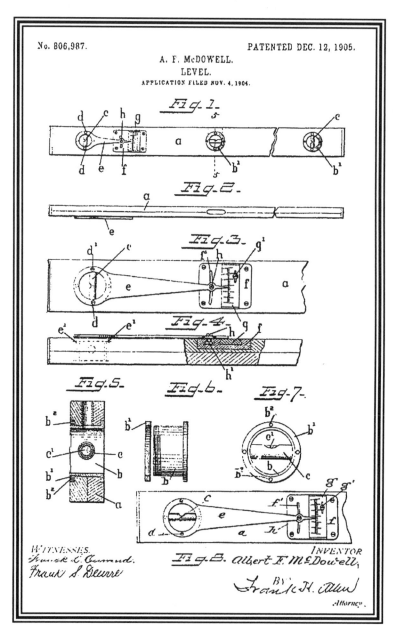

Assigned to the Stanley Rule and Level Company of New Britain, CT

LEVEL

For information on Justus A. Traut, see his patent of October 6, 1868 on page 28.

The operation of this adjustment is obvious from the patent drawings. The spirit vial casket is connected to the base plate via two machine screws that pass through springs, which facilitate the adjustment.

This is the interior adjustment for the level assembly adopted by Stanley in 1906, and used subsequently on the Victor series (Nºs 13, 14, 15, 19, etc.) and several other levels.

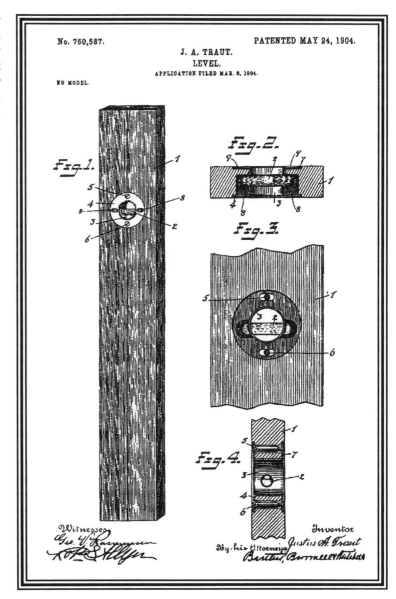

David Felix Broderick
Hartford, Connecticut

November 6, 1906
834,964

Assigned one-half to Lewis Sperry of Hartford, CT

TRY-SQUARE

A D. F. Broderick owned a utility contracting company. He may have become a plumber, but he was an apprentice in 1906-07.

Lewis Sperry was an attorney in the firm of Sperry and McLean. This was apparently a politically well-connected firm because Sperry had been a member of Congress from 1891-94, and McLean was Governor of Connecticut during 1903-04.

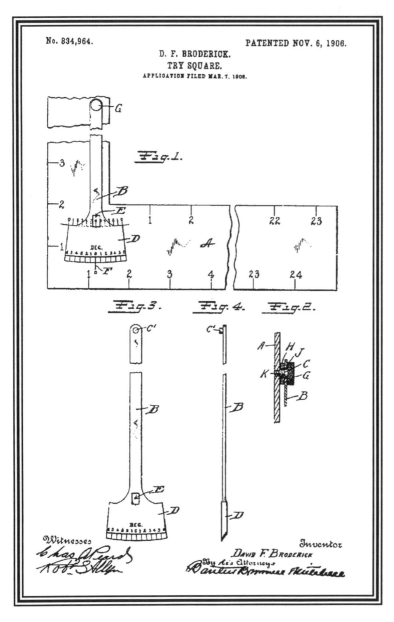

It is claimed that the patent is for an improvement in try-squares. The concept involves a fan-shaped plummet hanging on a framing square. The plummet's bottom edge is marked to read in degrees against an indicator mark. The left and right edges of the plummet can indicate against a scale on the body of the square.

No example of this device has been identified to date.

LEVEL

Beginning in 1903, Leonard Sheflott was listed in the New London, CT City Directory with an occupation of "stone mason." In 1906, he also had a business listing under "Masons and Builders."

This is a patent for a very simple inclinometer with an indicating needle, controlled by a suspended weight on the same shaft but in a compartment behind a graduated scale. The shaft or arbor is supported on a bearing surface in the front glass and in the rear of the case.

It is not known if this specific patent was used on a manufactured device. The general principal seems to flow through many subsequently patented inclinometers. This device could not be recognized unless it was signed or the patent date was stamped on it somewhere.

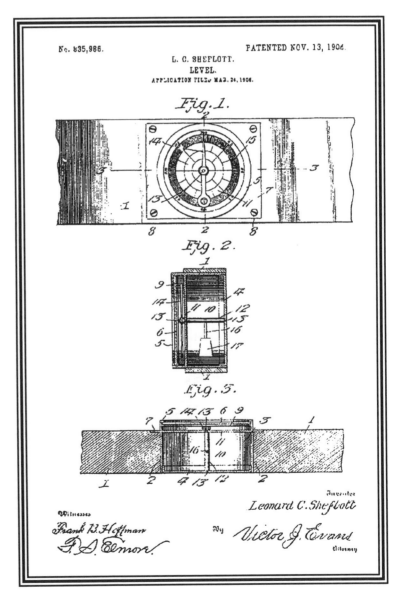

No. 835,986.

PATENTED NOV. 13, 1906.

L. C. SHEFLOTT.
LEVEL.
APPLICATION FILED MAR. 24, 1906.

Fig. 1.

Fig. 2.

Fig. 3.

Witnesses
Frank B. Hoffman
F. A. Elmor

Inventor
Leonard C. Sheflott
By Victor J. Evans
Attorney

LEVEL

According to Waterbury City Directories in 1908, Brown was a clerk at R. C. Co. (Randolph Clowes Co., Copper & Brass Rolling Mills) In 1909, he was manager of Risdon Tool Works. In 1910 he was clerk at Waterbury Brass Goods. In 1913 he was listed as being in "Insurance" and by 1930 he was said to be in real estate.

Hayden Brown's device is a level vial mounted on a sighting tube in a frame that swivels on a base. The base is supported on four screws that form adjustable feet.

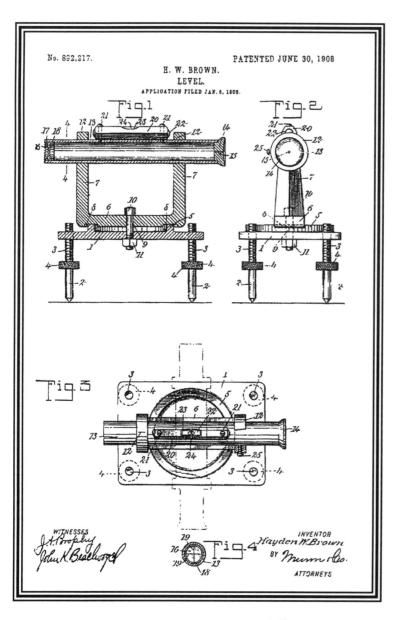

This is an exceedingly simple device with peephole and cross hairs. It seems amazing that this device could have been patented at all, let alone as late as 1908 with less than six months between the application filing and the granting of the patent.

As with some of the other patents, this one resembles a generic little sighting level. A device in the author's collection is quite similar, but contains a yoke underneath in place of the four adjustable legs. The yoke would allow mounting on a wooden carpenters' level. No example that strictly follows the patent description has been identified to date.

Justus A. Traut and Frank L. Traut
New Britain, Connecticut

November 3, 1908
902,778

Assigned to Stanley Rule and Level Company of New Britain, CT

PLUMB LEVEL

This was Justus Traut's last patent for a level. In a sense, it is fitting that his son, Frank L. Traut, was the co-patentee. For information on Justus A. Traut, see his patent of October 6, 1868 on page 28.

The patent describes a plumb fixture case with several tongues, or prongs, on the bottom. The prongs on the bottom of the case do a good job of holding the assembly in place. The patent also describes the construction of the case, and the adjustment feature on the top of the case.

The adjustment feature is a continuation of an earlier Traut patent. The top prongs, or lugs, as the patent refers to them, fold over each other when assembled. An adjustment screw passes through the slotted guide plate 5 and through the lugs which are threaded. Loosening the adjustment screw at the top end of the case allows the manual adjustment.

This was the plumb vial interior adjustment used in the Stanley № 3 and in the Victor series of levels.

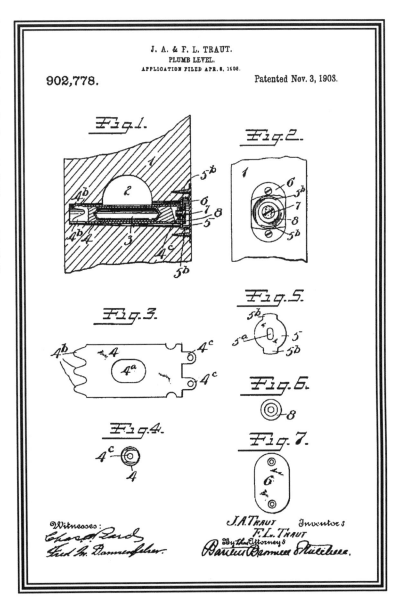

Harry H. Woolson
Medford, Massachusetts

January 5, 1909
909,046

TAILOR'S INDICATOR

Harry H. Woolson was a tailor, with a shop at 47 Winter Street, throughout the years from just before to just after this patent.

This device is intended to help determine the middle of the back when taking measurements for a garment. It provides the appropriate transverse measurements and indicates the amount by which one shoulder is higher than the other. Ruled arms move out simultaneously from either end of the horizontal cross piece. The ruled arms contain fingers that are intended to be inserted into the armpits of the customer. A curved spirit vial associated with a graduated scale (as shown on the first sheet of the drawings), or a pointed arm fixed at the top and indicating against a scale at the bottom (as shown on the second sheet of the drawings), are two suggested methods of determining the deviation from level of the shoulders. There are some obvious problems with this device in that the scale only represents a difference at a given rule extension.

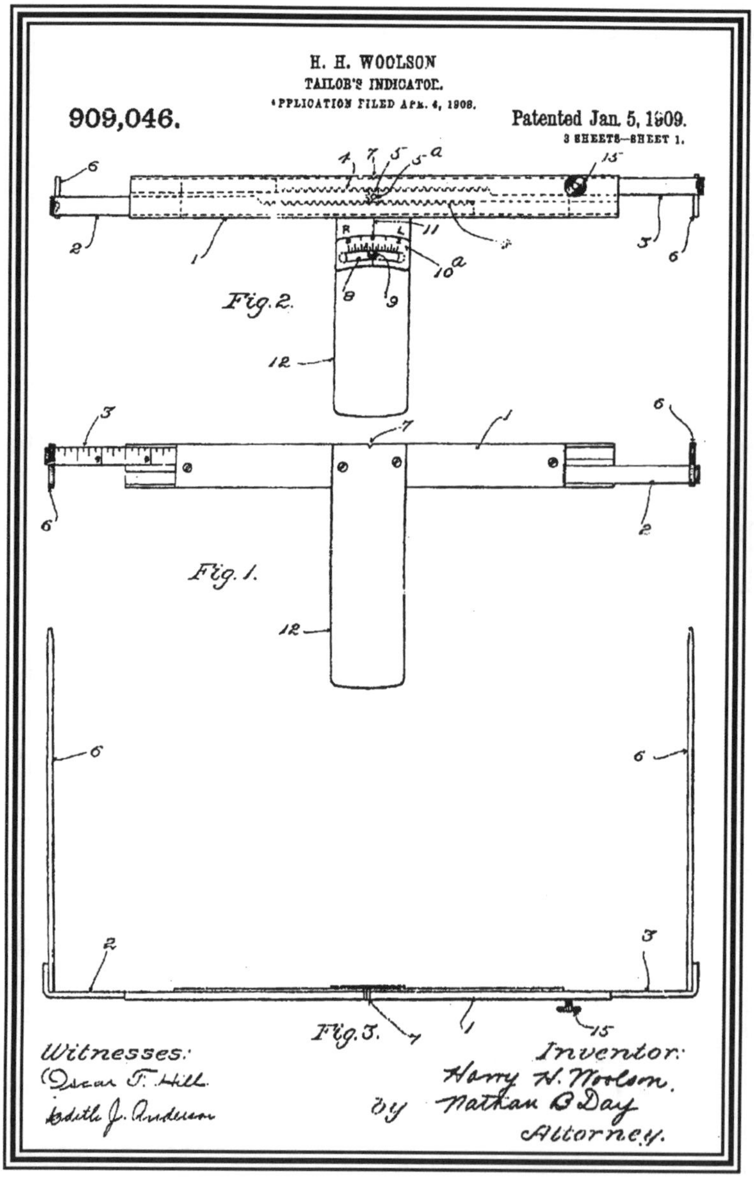

H. H. WOOLSON
TAILOR'S INDICATOR.
APPLICATION FILED APR. 4, 1908.
909,046.
Patented Jan. 5, 1909.
3 SHEETS—SHEET 1.
Fig. 2.
Fig. 1.
Fig. 3.
Witnesses:
Oscar F. Hill
Edith J. Anderson
Inventor:
Harry H. Woolson.
by Nathan B Day
Attorney.

There is no information to indicate that this device was ever produced.

PATENT DRAWINGS CONTINUED ON NEXT PAGE.

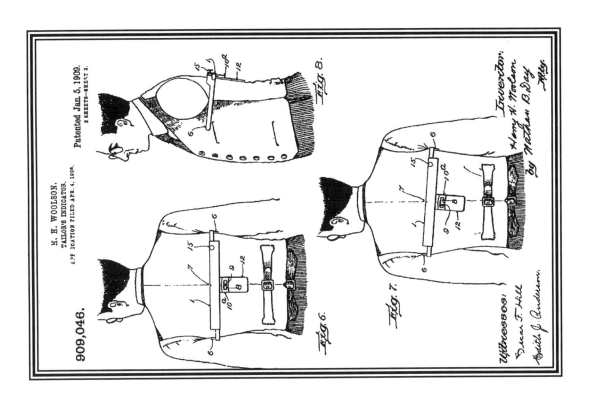

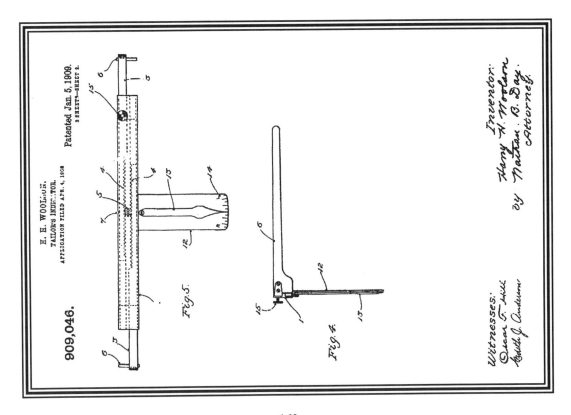

-169-

LEVEL

William J. Neidl had a long and varied career as a locksmith, machinist, inventor and entrepreneur. He was first found in the New Britain City Directories, in 1876, listed as a locksmith, living at 33 Linwood. By 1883-84, he was living at 55 Winthrop and was employed as a machinist at Russell & Erwin; by 1887, he had moved to a house at 81 Linwood. Although his address and employment didn't change, in 1894 his name was listed as Wengle J. Neidl. In 1895 he was listed as a manufacturer of ventilating sash locks with a business location at 497 Myrtle. In 1896 he was listed as being involved in hardware manufacturing in Meriden, but still living in New Britain. In 1898 he was a partner in W. J. Neidl and G. A. Hagist Hardware Manufacturing at 200 Elm St. From 1898 to 1901 he was employed by Landers, Frary & Clark (hardware and table cutlery manufacturers.); in 1902 he was employed by the Skinner Chuck Co. From 1903 to 1912 he was employed at the Stanley Works, during which period he obtained this patent. (Remember that at this time, The Stanley Works did not manufacture tools.) In 1915 and 1916 he was Vice-president of New Britain Hardware, which manufactured other levels patented by him. In 1917 he was listed without employment, but from 1918 to 1920 he was shown to be employed by Beaton & Caldwell, a steam heating specialties manufacturer.

This patent is for an all-metal, mantel clock-shaped inclinometer with a semicircular pendulum. The mantel clock portion contains a covered channel to guide the motion of the enclosed pendulum. Zero inclination marks on the pendulum can be read through the bottom of the channel. Slope can be read on the graduated face using indicators on the top of the pendulum. Sheet metal is recommended for construction.

A different sized level could be produced merely by adding a bottom of different length. This concept is carried into Neidl's later levels.

To date, there are no known examples of levels according to this patent.

PATENT DRAWING ON NEXT PAGE.

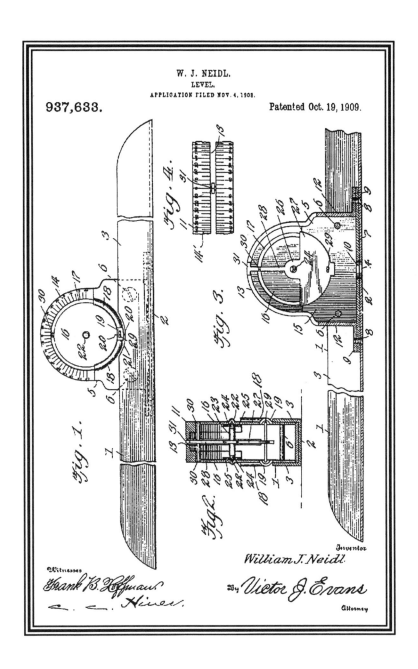

Inventor
William J. Neidl.

By Victor J. Evans
Attorney

Witnesses
Frank B. Hoffman
C. C. Hines

Napoleon R. Thibert
Worcester, Massachusetts

December 7, 1909
942,114

Assigned to Matthew J. Whittall and Alfred Thomas of Worcester, MA

PROTRACTOR AND LEVEL

Thibert was shown as a photographer in the 1903 Worcester City Directory. In 1904 and 1905, he was listed as a maker of cuspidors. In 1909, he was shown as a machinist.

In 1903, Alfred Thomas was engaged in yarn manufacture and Matthew Whittall was engaged in carpet manufacture (The Worcester City Directory of 1903 lists the business as Whittall and Thomas - Edgeworth Mills).

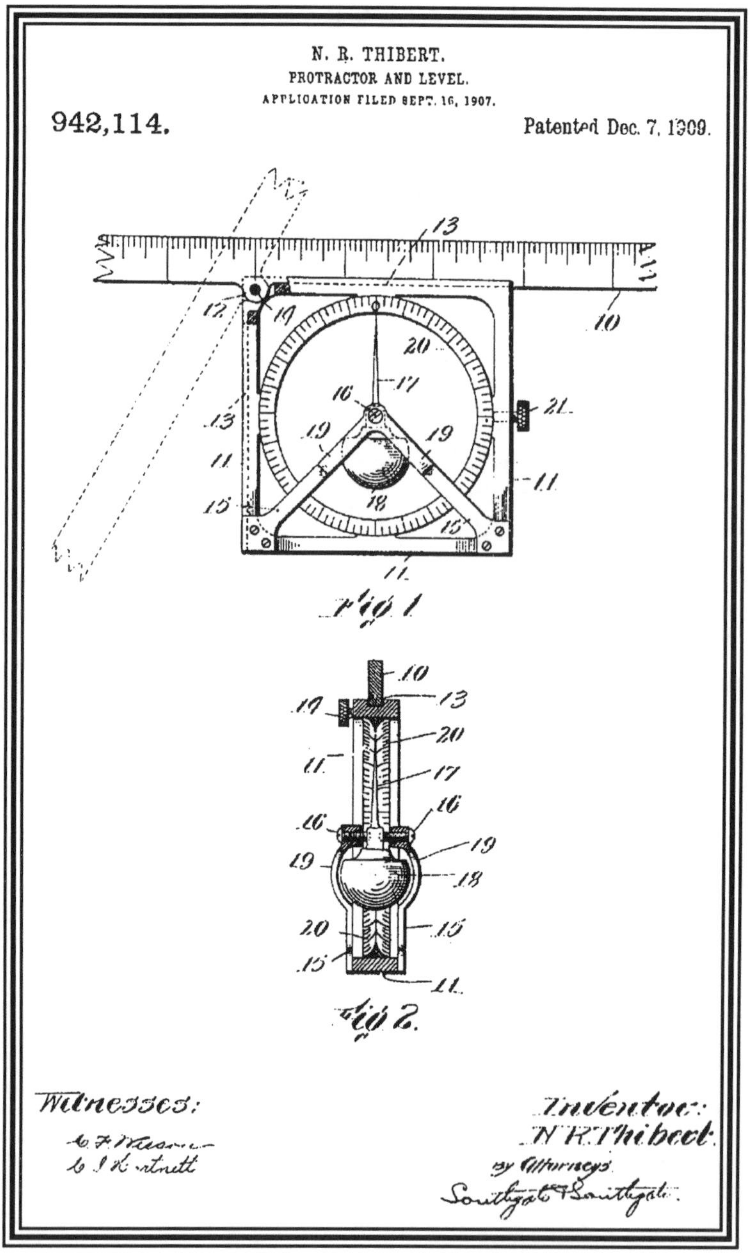

N. R. THIBERT.
PROTRACTOR AND LEVEL.
APPLICATION FILED SEPT. 16, 1907.

942,114.

Patented Dec. 7, 1909.

Fig 1.

Fig 2.

Witnesses:

Inventor:
N. R. Thibert
by Attorneys
Southgate & Southgate

This device consists of a counter-weighted needle *17* suspended inside a circular graduated frame *20* which is, itself, inside of a square frame *11*. One corner of the square frame is attached to a rule and is thus free to rotate 90°, about the pivot *14*. A knurled thumb screw acts as the pivoting axis and as a locking mechanism. The square frame is recessed on two sides to allow it to become square to the level.

The graduated circular frame may be rotated inside of the square frame for calibration. It pivots about the axis *16* and is locked into position by the knurled thumb screw *21*.

No example of such a device has been identified to date.

Assigned to the Stanley Rule and Level Company of New Britain, CT.

SPIRIT LEVEL

James Burdick was an Asst. Superintendent at the Stanley Rule & Level Company in 1910 and was General Superintendent by 1916.

For information about Christian Bodmer, see his patent of June 23, 1896 on page 121.

This patent describes a vial casing that could be used as either a plumb or a level fixture. Four light-weight anchors, shown in *Fig.5,* fit into the porthole and can be used to support the vial casing in an adjustable manner. The vial case, shown in *Fig.4*, is to be made from a sheet metal blank shaped as shown in *Fig.3*.

It is not clear that this device was ever used by Stanley. It was, however, similar in style to the type of vial casings being patented by others and employed by others. The general idea behind this vial casing is that it was easily and economically formed out of a single piece of light-weight brass.

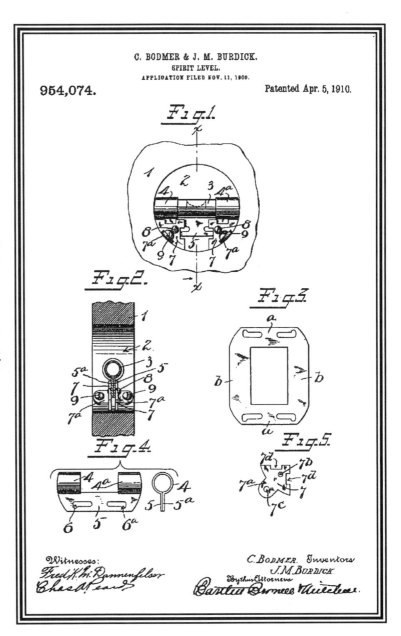

C. BODMER & J. M. BURDICK.
SPIRIT LEVEL.
APPLICATION FILED NOV. 11, 1909.

954,074.

Patented Apr. 5, 1910.

Assigned to Stanley Rule and Level Company of New Britain, CT

SPIRIT LEVEL

For information about Christian Bodmer, see his patent of June 23, 1896 on page 121.

This patent is for a metal level stock to take vial units similar to those in № 954,074. The concept is similar to others being proposed at the time.

There is no evidence that this patent was ever applied by Stanley to a manufactured level. Application was made for this patent on the same day as for Patent № 954,074.

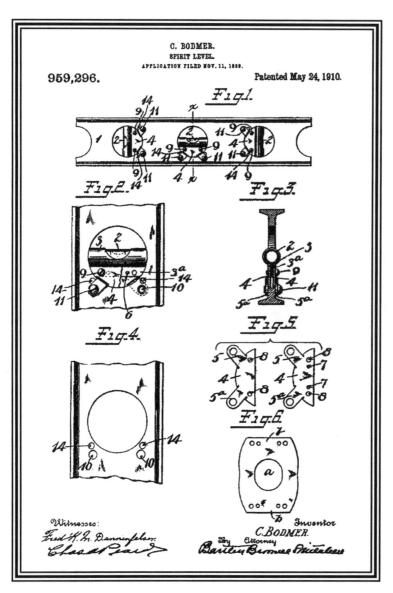

C. BODMER.
SPIRIT LEVEL.
APPLICATION FILED NOV. 11, 1909.

959,296.

Patented May 24, 1910.

CENTERING DEVICE

Alcide Franchini first appears in the 1909 New Haven City Directory and he was said to have worked for the H. C. Rowe Company, who were oyster growers and shippers. His name was given as Alceda. In 1911-12, Franchini was listed as a laborer but his name was given as Arcide. In 1915, Alcide Franchini was listed as an employee of Winchester Repeating Arms Co.

This patent describes a device that stands U-shaped on two spring-loaded legs. This tool is meant to be a center marker whose primary function is to mark the exact top and exact bottom of a shaft or pipe.

A spirit vial extends from side to side near the center of the device. The spirit vial must be straight but of greater diameter in the center than on the ends. In such a case, the vial can be read no matter which side of the vial is on top and the tool can be used upside-down. This vial is quite novel.

There is no example of this device known to the author and, although employment as a standard level was not its claimed use, it could function in such a capacity.

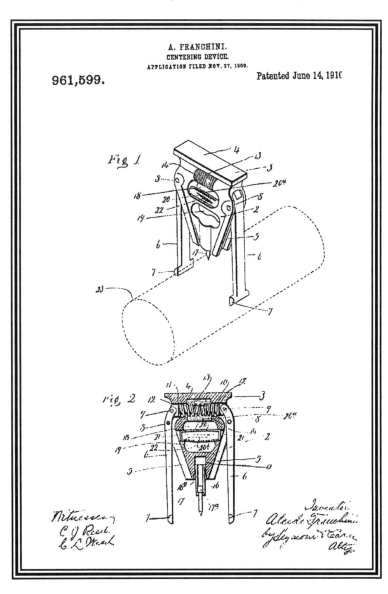

A. FRANCHINI.
CENTERING DEVICE.
APPLICATION FILED NOV. 27, 1909.

961,599.

Patented June 14, 1910

Fig 1

Fig 2

GRADE-LEVEL

Joseph E. DeBisschop was listed in the New Britain City Directory as being employed by H. P. Battey, a milk dealer, in 1910. There was no other listing for him either before or after 1910.

This patent is for a fairly complex grade level. The level is designed to be equipped with sights, with an indicating dial and pointer attached to a separate geared arm. The geared arm can be moved by means of an adjusting screw running through the length of the level to an end or porthole. There is a level vial at the base of the geared arm.

The pendulum is not free swinging, but is pinned to an arm that can be used to adjust its position. The use of gearing (the ratios are never stated) allows the user to determine grades as great as 100° different from the base of the level. Thus, the user sets the level on the surface he wants to use for reference and sights on the target. He then brings the spirit vial on the pendulum to a level position, using the adjusting screw on the end of the arm. The pointer attached to the inner geared wheel, will then indicate the angle between the reference surface and the target.

To date, no examples of a level built according to this patent are known.

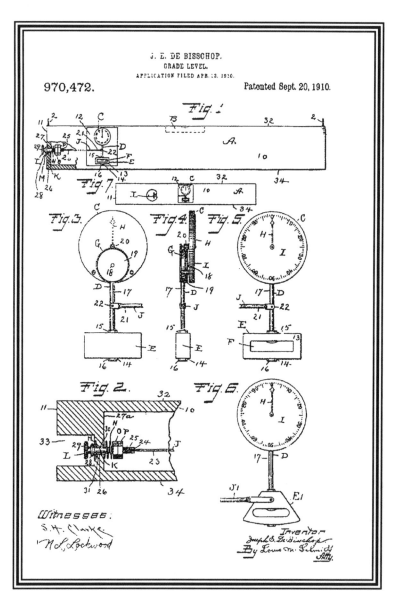

Dennis E. Reilly
Derby, Connecticut

October 4, 1910
971,792

LEVEL

Dennis Reilly was listed in the Derby City Directories, in 1908, as having the business, D. E. Reilly & Son, "Architects, Carpenters & Builders." But Daniel, the son, went to Boston and, in subsequent years, Dennis was simply listed as a carpenter. (In 1912, Daniel appeared again as Reilly Bros. & Co., sheet metal work, roofers, cornices and skylights.

The patent describes an attractive looking inclinometer device. A stylized triangular device, containing three wheels with flanges, rides around in a circular scale rail. A pointer is placed at one point of the triangle while a weight is added opposite that point. The classic plummet is replaced by a rolling weight in a track; and the problem introduced by shafts and bearings is avoided for the main weight, and reduced to one of operable axles for the wheels and a smooth track. However, this device would seem to be very vulnerable to dirt on the track or in the groove of the wheel.

No example of this device has been identified to date.

PATENT DRAWINGS ON NEXT PAGE.

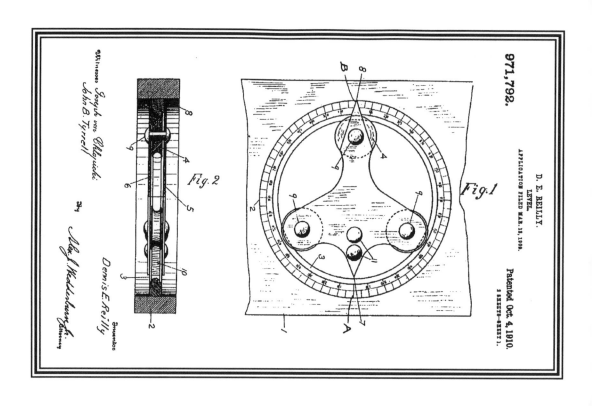

971,792.

D. E. REILLY.
LEVEL.
APPLICATION FILED MAR. 18, 1909.

Patented Oct. 4, 1910.
1 SHEET—SHEET 1.

Fig.1

Fig.2

Inventor
Dennis E. Reilly

Witnesses
Joseph von Oblazinski
John B. Tyrrell

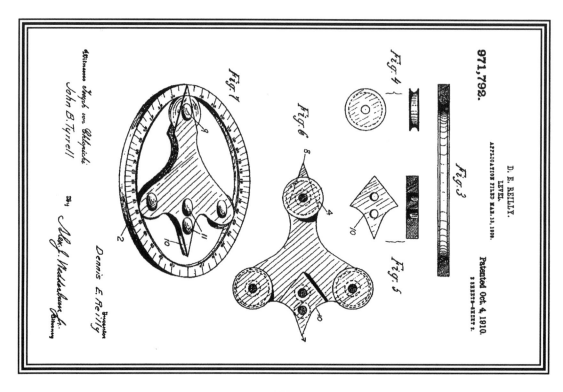

971,792.

D. E. REILLY.
LEVEL.
APPLICATION FILED MAR. 18, 1909.

Patented Oct. 4, 1910.
2 SHEETS—SHEET 2.

Fig.4

Fig.3

Fig.6

Fig.5

Fig.7

Witnesses
Joseph von Oblazinski
John B. Tyrrell

Inventor
Dennis E. Reilly

COMBINED SQUARE AND BEVEL PROTRACTOR

For information on Laroy Starrett, see his patent of May 6, 1879 on page 47.

This patent is claimed to be for improvements on Starrett's patent of December 27, 1904. The tool contains two opposing spirit levels in a rectangular stock. There is a graduated rotating piece at one end that receives the metal rule. The rule can be locked into place by a hook mechanism, as in similar Starrett tools. A recess, running the length of the stock, is provided on two sides to receive the rule in such a fashion that the outer edges of the rule and the stock are parallel. Adjusting studs are provided appropriately in the frame for the parallelism adjustment.

This is the patent for Starrett's № 439 "Builders Combination Tool".

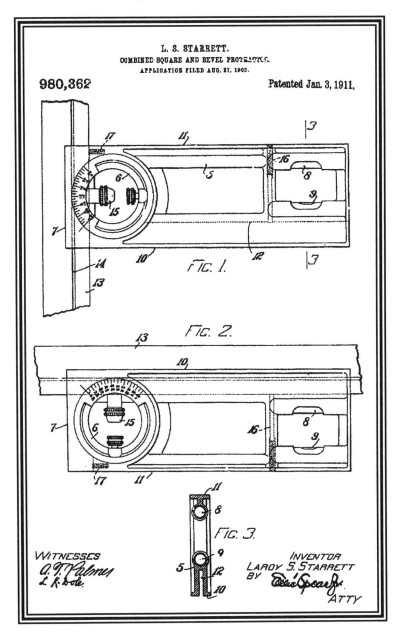

L. S. STARRETT.
COMBINED SQUARE AND BEVEL PROTRACTOR.
APPLICATION FILED AUG. 27, 1909.

980,362

Patented Jan. 3, 1911.

FIG. 1.

FIG. 2.

FIG. 3.

WITNESSES
O. T. Palmer
L. K. Dole.

INVENTOR
LAROY S. STARRETT
BY
Charles Spear
ATTY

LEVEL

No information about Daniel Lawrence was located. He was not counted in the 1910 Census.

This is a very simple device considering the date. It is an inclinometer of the weighted needle type, wherein the needle has a circular weight on the bottom. The axle is a short stud. The entire face is covered by a glass plate. The apparatus is excruciatingly simple and probably very inexpensive to manufacture. Its accuracy will be limited.

Although cheap and simple inclinometers are known, there is nothing to link the known items to this patent.

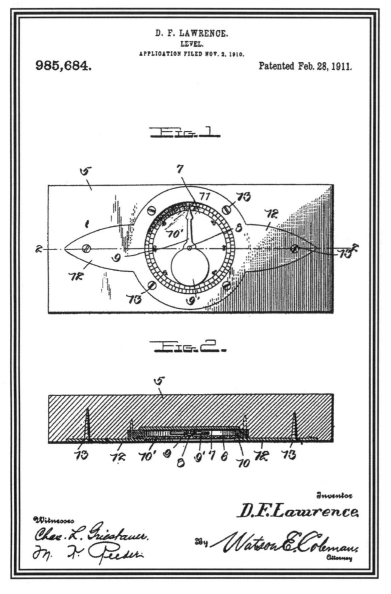

Alexander Johnson
Cromwell, Connecticut

May 2, 1911
991,446

PLUMB-LINE

No information about Alexander Johnson was found during the research for this book.

This patent describes a mason's level consisting of a hook with attached line, a spacer to keep the line off the wall and a two-pronged hook to carry the level cartridge.

There is no information to indicate whether or not this device was ever manufactured. Because of its nature, the chance of finding one intact seems remote.

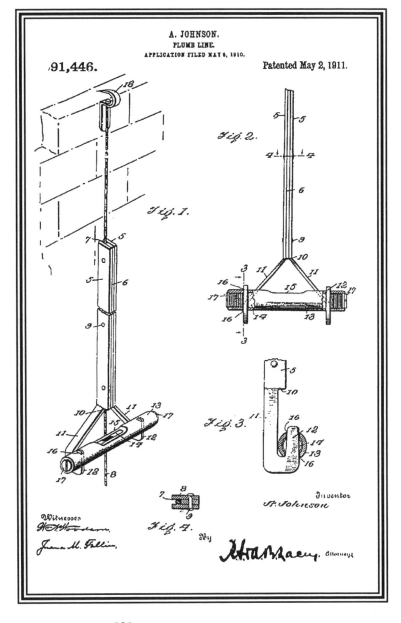

Daniel J. Seymour
Providence, Rhode Island

June 13, 1911
995,099

Assigned one-half to John T. Cuddy of Providence, RI

GRAVITY-LEVEL

The 1911 Providence City Directories listed Daniel Seymour as being in "tanks." The only listing under tanks was A. T. Stearns Lumber Co. that made wooden tanks. In 1910 Seymour had been listed as being a Commission Merchant.

John T. Cuddy was listed as a draftsman. Both men continued at these general occupations for several years.

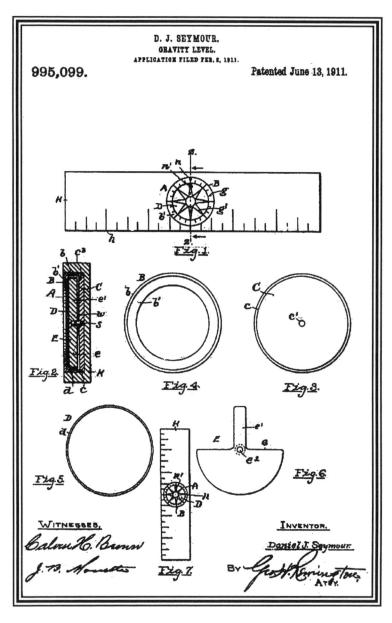

D. J. SEYMOUR.
GRAVITY LEVEL.
APPLICATION FILED FEB. 2, 1911.

995,099.

Patented June 13, 1911.

This patent describes an inclinometer of the sector type. It utilizes a semi-circular weight with an upright finger. The weight is attached, on an axle, to a rotatable dial or, through a dial, to a moveable indicator. Regardless of which is employed, the device is graduated on the annular outer casing, and an eight-pointed star shape is suggested for the indicator, or for a design on the dial.

Much attention is devoted to the method of making this device cheaply.

This device was apparently intended to be relatively inexpensive because "quite thin sheet-metal stock" was indicated to be the material of choice for the metal parts of its construction. It was also suggested that both edges of the wooden stock be ruled and that, perhaps, advertisements could be stamped on the wooden sides. No example of this device has been identified to date.

Laroy S. Starrett
Athol, Massachusetts

December 12, 1911
1,011,262

COMBINATION TOOL

For information on Laroy Starrett, see his patent of May 6, 1879 on page 47.

Starrett's patent describes this tool as fulfilling the functions of back square, plumb, level, mitre angle, circle scriber and beam gage. It may be used with one or two of the stocks. The tool resembles a combination square with the obvious difference that the stock is clamped to the blade with a clamp-piece *4* actuated by the thumb screw *3*. With two of the stocks, a thickness gage for beams can be made. Each stock contains a spirit vial and carries an awl. Holes are provided at *13*, so that, using the awl as a pivot, circles may be struck having radii as great as the rule. Like most Starrett patents, the functions are obvious from the patent drawings.

The author is unaware of any production of this device.

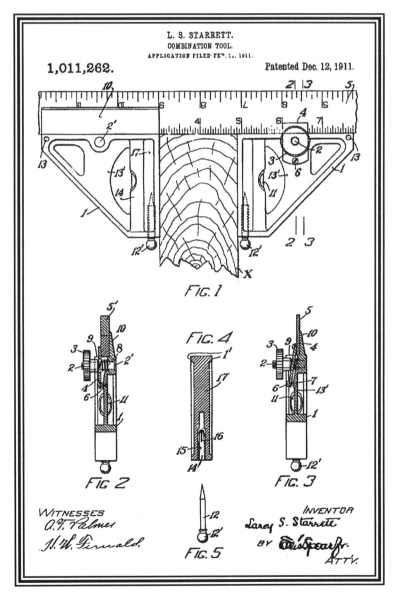

Albert Makowski
Meriden, Connecticut

SPIRIT-LEVEL

Makowski was an Austrian citizen apparently living in or around Meriden CT at the time that the patent was granted. No other information about Makowski could be located.

The object of this patent is to provide a standard carpenters', or masons', level with a set of brackets by each vial and a removable "flashlight" that can be held in the brackets as desired. The on switch is said to be activated by a slight displacement of the light to the side containing the switch (on the left side in the drawings) and it is said to remain in this position without further assist. The light is turned off by a slight movement in the opposite direction.

When not in use, the light is stored in the end of the stock.

No example of a level constructed according to this patent has been located to date.

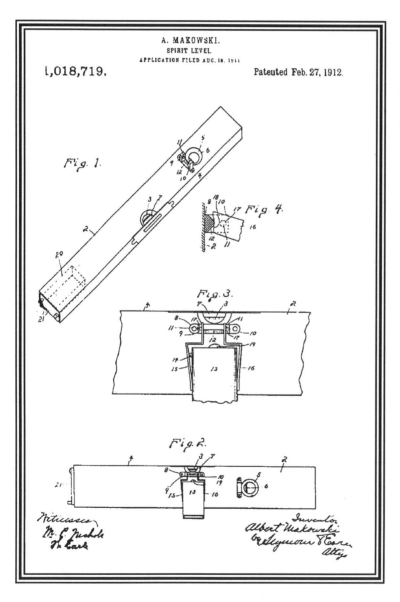

A. MAKOWSKI.
SPIRIT LEVEL.
APPLICATION FILED AUG. 18, 1911.

1,018,719.

Patented Feb. 27, 1912.

Fig. 1.

Fig. 4.

Fig. 3.

Fig. 2.

James M. Burdick and Edmund A. Schade
New Britain, Connecticut

January 14, 1913
1,050,610

Assigned to the Stanley Rule and Level Company of New Britain, CT

LEVEL

According to Roger Smith[21], Edmund Schade was born in Germany in 1855, joined Stanley in 1873, and rose through the ranks to become Mechanical Superintendent in 1900, a position that he still held at the time of his death in 1932.

For information about James M. Burdick, see his patent of April 5, 1910 on page 173.

In this patent, a bottom plate is fastened to the bottom of the mortise and the vial casket is fastened to the bottom plate with machine screws. The left tongue of the vial casket is to be formed into a bent spring in order to keep tension on the screws.

This patent is claimed by Walter[22] to be the adjustment mechanism used in Nº 1193 and others. If this is correct, then the use is not faithful to the patent because, in the Nº 1193, a spiral spring is used under the tongue on the left side of the vial casket.

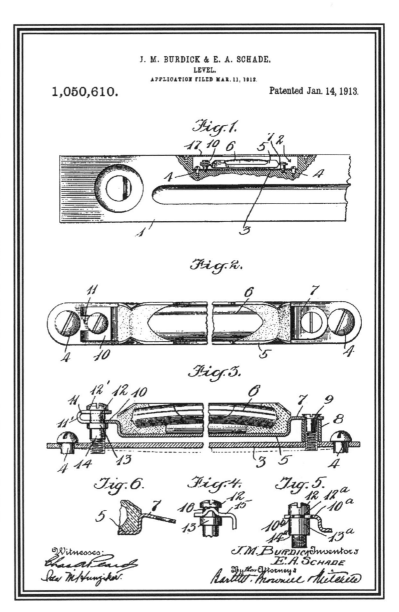

J. M. BURDICK & E. A. SCHADE.
LEVEL.
APPLICATION FILED MAR. 11, 1912.

1,050,610.

Patented Jan. 14, 1913.

Fig. 1.

Fig. 2.

Fig. 3.

Fig. 6.   Fig. 4.   Fig. 5.

Witnesses:

J. M. Burdick Inventors
E. A. Schade
By their Attorneys

21   Smith, Roger K., *Patented Transitional and Metallic Planes in America, Vol II*, Roger K. Smith Publisher, Athol, MA 1992, pp 224-228.

22   Walter, John, *Antique & Collectible Stanley Tools, A Guide to Identity & Value, 2ⁿᵈ Ed.* Published by The Tool Merchant, Marietta OH 1996, p 785.

John Scott and Nicholas McGrath
Southington, Connecticut

May 13, 1913
1,061,638

SPIRIT-LEVEL

Both Scott and McGrath were longtime Southington employees.

The patent calls for placing plates in the bottom of the mortises for the vial casings. The plates are to have a recessed place wherein a roller bearing is inserted. The level vial casket has a recess on the bottom that fits over the roller bearing. The adjustment is made through two screws in the top of the top plate. The plumb fitting is attached to the plumb vial cover plate, and passes through the plate in the bottom of the mortise. The plumb vial cover plate sits on another roller bearing, and the two screws in the top plate pass through the bottom plate. Turning the screws that hold the plumb vial cover plate onto the level will adjust the plumb vial setting.

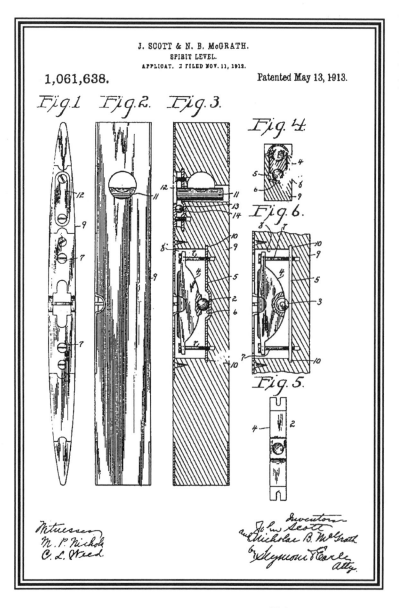

This is the patent used for Southington's torpedo level and the product was faithful to the stipulations of the patent.

Per Erik Ekman
Woburn, Massachusetts

June 3, 1913
1,063,642

COMBINATION CENTERHEAD, LEVEL, AND BEVEL-PROTRACTOR

Per Erik Ekman was a machinist who, along with seven siblings, boarded at his father's home in Woburn.

The operation of the patented device is obvious from the drawings in that the centerhead is placed on the shaft and, using the rotating protractor, the inclination of the instrument can be determined. Thus, the device qualifies as a shaft level. Alternatively, the pivoting spirit vial case can be set, and the inclination can be read from the scale on the protractor head. A prick punch is set into the center head. In *Fig.1,* the tool can be used to find top dead center of a shaft. *Fig.2* shows how the device can be used to locate and mark a position at any angle from top center. Both the spirit level and the protractor head are adjustable. The level is adjusted using a geared mechanism at *31,* while the protractor is adjusted manually when the thumb screw *5* is loosened. *Figs.3* and *4* show how angles may be transferred from a plane surface to a shaft using the level and the protractor head.

It is perhaps only marginally acceptable to classify this tool as being useful as a level. However, the inventor calls it a level. To date, no examples of a device, made according to this patent, are available.

PATENT DRAWINGS ON NEXT PAGE.

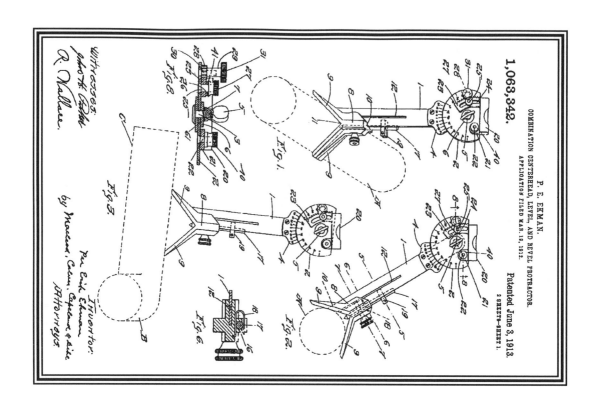

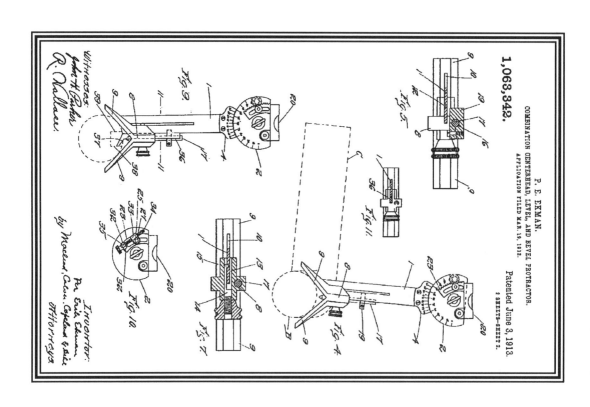

Edmund A. Schade
New Britain, Connecticut

September 22, 1914
1,111,677

Assigned to the Stanley Rule and Level Company of New Britain, CT

HAND LEVEL CONSTRUCTION

For information about Edmund A. Schade, see his patent of January 14, 1913 on page 185.

This is the "Stanleyization" of the McNutt patent. It looks very little different from the original, but the pointed knurled nut has been replaced by a slotted knurled nut and the sleeve body is threaded rather than pinned.

The concepts of this patent along with the McNutt patent (McNutt was from Arkansas) were used in Stanley № 137 millwrights' level.

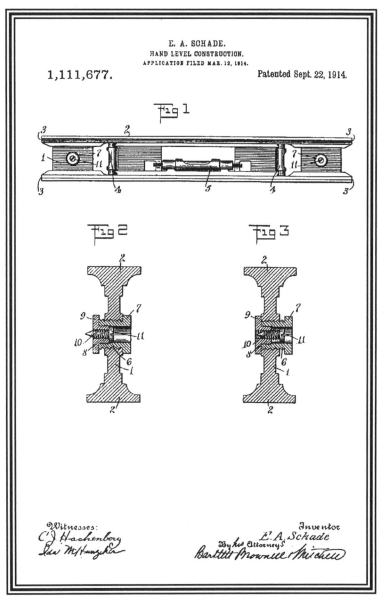

George W. Bartlett
Meredith, New Hampshire

January 12, 1915
1,124,833

TIMBER MEASURING DEVICE

At this time, no information is available on George Bartlett.

In its intended purpose, this device can be employed to measure the amount of lumber in a standing tree. The scales *9* and *10* are developed for use at a certain distance from the base of a tree. The front sight contains two vertical wires that can be used to determine the diameter of the tree, and the sight itself can be widened by insertion of a wedge.

Its use as a sighting level is obvious and with a proper scale it can be used to set elevations.

No example of this device has been identified to date.

PATENT DRAWINGS ON NEXT PAGE.

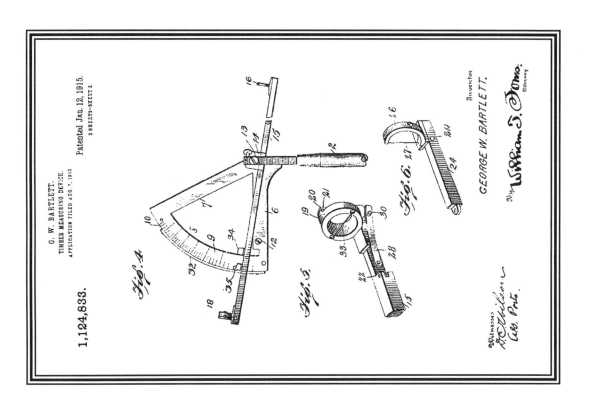

G. W. BARTLETT.
TIMBER MEASURING DEVICE.
APPLICATION FILED AUG. 7, 1913.

1,124,833.

Patented Jan. 12, 1915.
2 SHEETS—SHEET 2

GEORGE W. BARTLETT,

Inventor

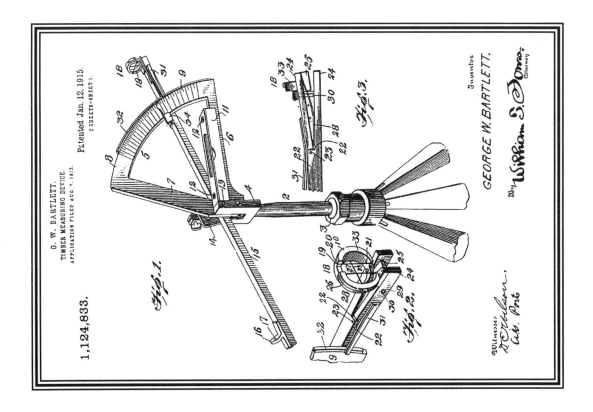

G. W. BARTLETT.
TIMBER MEASURING DEVICE.
APPLICATION FILED AUG. 7, 1913.

1,124,833.

Patented Jan. 12, 1915.
2 SHEETS—SHEET 1

GEORGE W. BARTLETT,

Inventor

Matti Hamalainen
Ludlow, Vermont

June 8, 1915
1,142,418

COMBINATION TOOL

No information was located about Matti Hamalainen.

This describes a tool having two side plates and a central longitudinal recess. Two additional arms are to be pivoted inside of the side plates, one at each end. The side plates and arms are ruled. The ends of the arms contain several notches to allow the arms to be locked in various positions. The central stock carries spirit vials in both the plumb and level positions.

To date, no examples of this device have been identified.

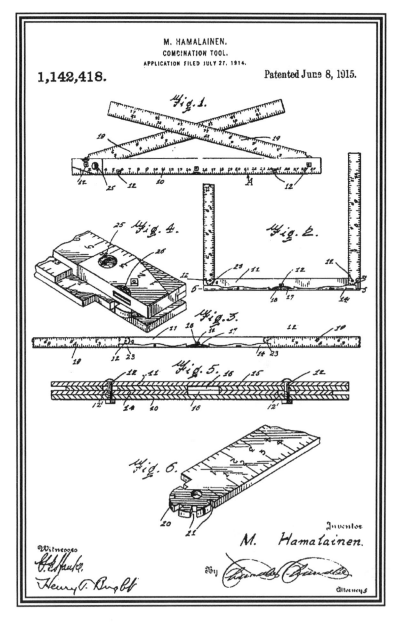

Edmund A Schade
New Britain, Connecticut

August 24, 1915
1,151,549

Assigned to the Stanley Rule and Level Company of New Britain, CT

LEVELING STAND AND BASE

For information about Edmund A Schade, see his patent of January 14, 1913 on page 173.

The patent describes a two piece unit. The upper member of the unit contains a means of clamping a level in place, adjustable feet, and a central combined pivot and clamping screw. The lower unit provides a platform, or table, on which to rest the turntable, and a socket for attaching the combined unit to a post or staff.

The concepts of this patent were embodied in the Stanley leveling stands Nºs 38, 48, and 338. The products faithfully represented the patent.

PATENT DRAWINGS ON NEXT PAGE.

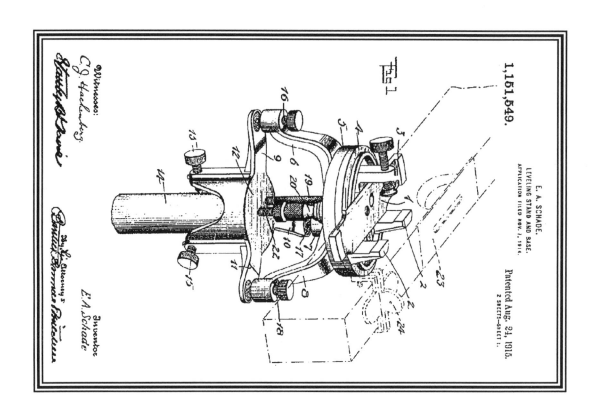

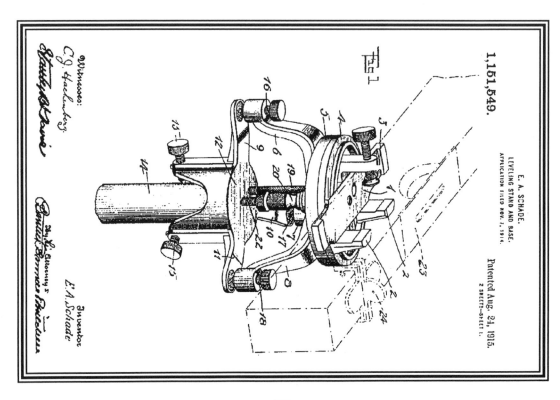

Frank B. Miller
Unionville, Connecticut

November 9, 1915
1,159,522

Assigned to Union Cutlery and Hardware Co. of Unionville, CT

LEVEL

In1907 Frank Miller was employed as a machinist at Union Cutlery in Unionville, CT. The Unionville City Directory also listed him as a watchmaker. (He was listed in the business directory as a watchmaker.) In the 1910 census Mr. Miller was reported to be employed as a "superintendent cutlery." In the census of 1920 he was reported to be employed as "Foreman hardware." These may have been the same job or have represented a transfer to another department within Union Cutlery and Hardware Company.

The patent was for a level formed from a single piece of sheet metal. The object of the patent was for a device that was simple, neat, and inexpensive.

Examples of this level are known, although they are often overlooked by collectors because of the simple and inexpensive nature of the tool.

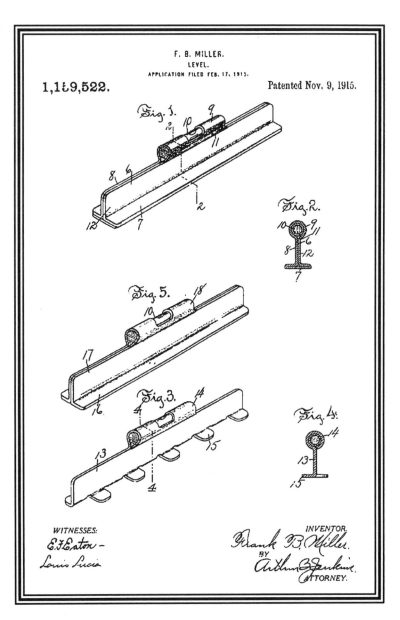

William Josep Neidl
New Britain, Connecticut

March 28, 1916
1,177,131

Assigned to New Britain Hardware Manufacturing Company of New Britain, CT

LEVEL

Neidl was Vice President of New Britain Hardware, which was one of the many ventures in which this entrepreneur was involved during his life. (See p. 170.)

The patent concepts involve a removable cover for the body, an internal metal frame, and three level cases, fixed at one end and adjustable by an externally protruding screw at the other. The level is entirely contained in the body and can be installed on a flat base of any length. A part of the concept is that even long levels can be light-weight. Construction is to be of sheet metal of such thickness as to be easily folded. The base of the level is to be formed as a channel into which the cover for the body can be inserted. The internal frame can be attached to the center of the channel.

The level was manufactured by New Britain Hardware in several sizes, from 4½" to 24", all employing the same body on a base of differing lengths. This concept had been introduced in his earlier patent. The levels were faithful to the patent concepts.

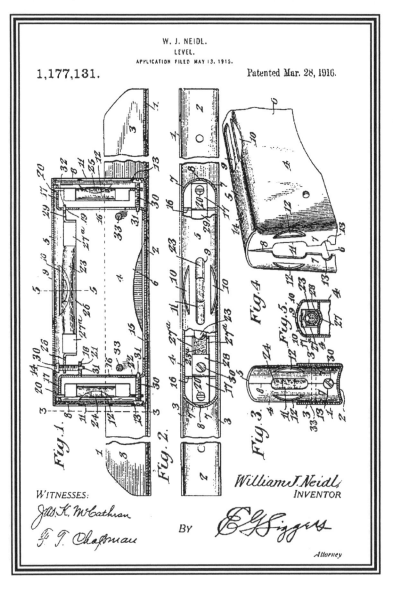

W. J. NEIDL.
LEVEL.
APPLICATION FILED MAY 13, 1915.

1,177,131.

Patented Mar. 28, 1916.

Fig. 1. Fig. 2. Fig. 3. Fig. 4. Fig. 5.

WITNESSES:
Jas. K. McCathran
Geo. T. Chapman

William J. Neidl
INVENTOR

BY E. G. Siggers
Attorney

ANGLE-GUIDE FOR TOOLS

Louis Arkin was a physician in Boston.

The device is constructed with two planes at right angles to each other. One plane contains a spirit level oriented perpendicularly to the second plane. The other plane is semi-circular and contains a spirit vial that can be manually positioned at a desired angle. The edge of the semi-circle is graduated from 0° - 180°. When the adjustable level is placed at 0°, the two levels assume a cross test appearance. In the same plane as the fixed level there is to be another surface part whose function is to provide attachment to the tool by unspecified means.

No such device has been identified to date. The device is proposed for use in conjunction with any tool that must be operated at a predetermined angle, and is not restricted to hand drills.

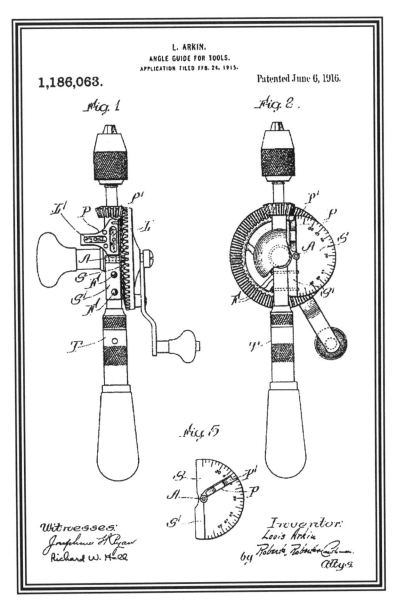

L. ARKIN.
ANGLE GUIDE FOR TOOLS.
APPLICATION FILED FEB. 24, 1915.

1,186,063.

Patented June 6, 1916.

Fig. 1

Fig. 2

Fig. 5

Witnesses:
Josephine S Ryan
Richard W. Hall

Inventor:
Louis Arkin
by Roberts, Robertson...
Atty's

Assigned to the Stanley Rule & Level Company of New Britain, CT

SIGHTING ATTACHMENT FOR LEVELS

For information about Christian Bodmer, see his patent of June 23, 1896 on page 121.

This patent describes sights that may be clamped to any standard level stock. There are several changes between the patent designs and the manufactured product. *Figs. 1, 2,* and *3* clearly show the anticipated means of securing the sight to the stock. In practice, this became a base with two screw holes on each side, and very long clamping screws so that the screws would be able to reach all the way to the rail, or web, on a metal stock. Among the main features of the patent, and one that was retained, is the use of "lugs" struck from the frame of the front sight to hold the crossed wires. This should have been a feature that was easy to manufacture, in that the frame needed only to be struck hard, with some kind of punch, to lock in the cross wires.

The patent envisions machined flanges on the base. This was not incorporated. The patent envisioned annular flanges that held the crossed wires and peep hole slug. These were not incorporated. The changes that were made probably were all made in order to lower the cost.

Bodmer's patent was applied to the Stanley № 138 level sights which worked on wood or metal levels.

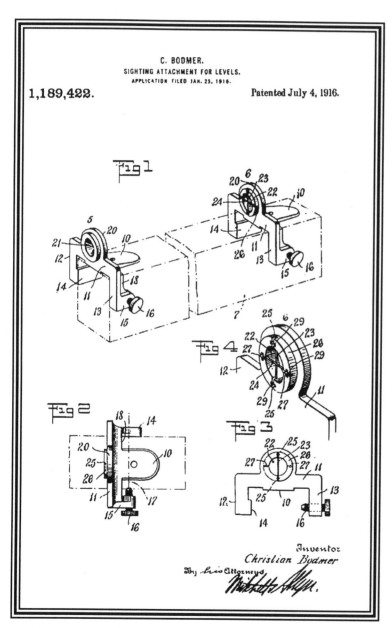

C. BODMER.
SIGHTING ATTACHMENT FOR LEVELS.
APPLICATION FILED JAN. 25, 1916.

1,189,422.

Patented July 4, 1916.

Inventor
Christian Bodmer
By his Attorneys

COMBINED GAGE AND LEVEL

For information on Laroy Starrett, see his patent of May 6, 1879 on page 47.

This patent describes a device that can be used as a gage for hard to access surfaces. A spirit level vial set in the base allows the device to function as a level. Actual distance measurements are made with a micrometer or similar device, usually for determining the distance between surfaces *3* and either *10* or *11*.

This is the patent for Starrett's "Planer and Shaper Gage" № 246.

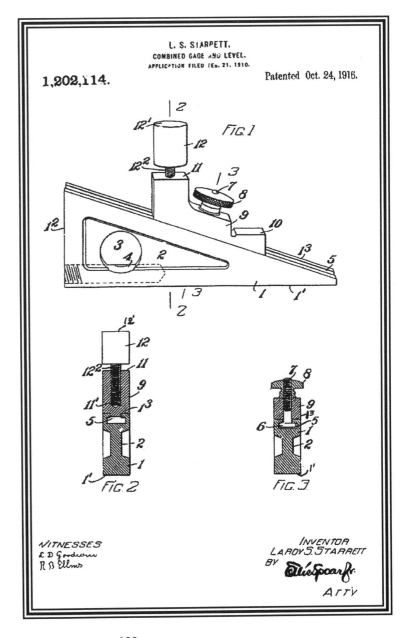

George W. Lyons
Orange, Connecticut

November 21, 1916
1,205,946

Assigned three-fourths to Abraham L., James A. and Matthew A. Notkins of New Haven, CT.

LEVEL

The New Haven City Directory for 1916 lists Lyons as Manager of the New England Iron Works, a company that dealt in ornamental and structural iron.

A. L. Notkins & Sons were dealers in real estate & insurance, as well as being builders. But James A. Notkins was also listed as a druggist, James A. Notkins & Bro. The brother in this case is Benjamin Notkins.

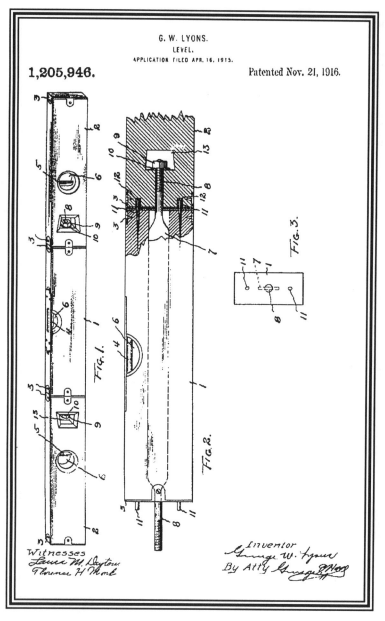

G. W. LYONS.
LEVEL.
APPLICATION FILED APR. 16, 1915.

1,205,946.

Patented Nov. 21, 1916.

FIG. 1.
FIG. 2.
FIG. 3.

Witnesses
Inventor
George W. Lyons
By Att'y

This patent is for a breakdown level utilizing a metal bar, threaded on the ends and imbedded in the center section of a three-section level. The center section also has two dowel pins extending out from each end in order to register the pieces when assembling them. The threaded portion of the metal bar extends through a hole in the end pieces, and a nut is attached at a port hole to hold the pieces firmly together.

To date, no examples of this level have been located.

Edmund A. Schade
New Britain, Connecticut

January 9, 1917
1,211,882

Assigned to the Stanley Rule and Level Company of New Britain, CT

SPIRIT-LEVEL

For information about Edmund A. Schade, see his patent of January 14, 1913 on page 173.

The patent describes the stud and base arrangement shown clearly in *Fig. 2* of the patent drawing. The base contains a threaded hole into which a stiff spiral spring is inserted. A stud containing a receptacle for the top of the spring is screwed into the hole. The end of the level cartridge is attached to the top of the stud. The stud also contains a hole so that a "pointed tool or nail" may be inserted to assist in turning the stud. Obviously, when the stud is turned, that end of the level is moved up or down.

This patent was used by Stanley for an improved adjustment on the № 34 machinist level.

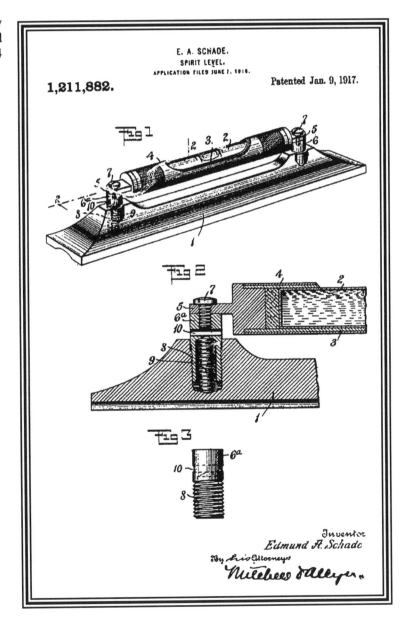

Christian Bodmer and Albert W. Ritter
New Britain, Connecticut

January 16, 1917
1,212,735

Assigned to Stanley Rule & Level Company.

SPIRIT LEVEL

Albert Ritter was an Assistant Superintendent at Stanley Rule & Level Company, at least during the period 1917-1924. He moved to New Haven in 1931.

For information about Christian Bodmer, see his patent of June 23, 1896 on page 121.

This patent describes a level stock and vial units like those used in the Stanley № 23. The stock is to be bored and counterbored, leaving a web in the center, a portion of which is removed on opposite sides of the boring. An annular fitting, with hollow bosses to receive the spirit tubes, is placed in the hole. The annular fitting has a flange with elongated holes to allow for rotary movement of said flange. When the flange rotates the spirit vials will rotate with it, and the level can be adjusted. The fittings are covered with a cover glass held in place by a suitable material, such as putty. Split fittings are to be employed when more than one spirit vial is destined for the same boring as in *Fig.6*.

This patent was applied in the production of Stanley's № 23, 23S and other levels.

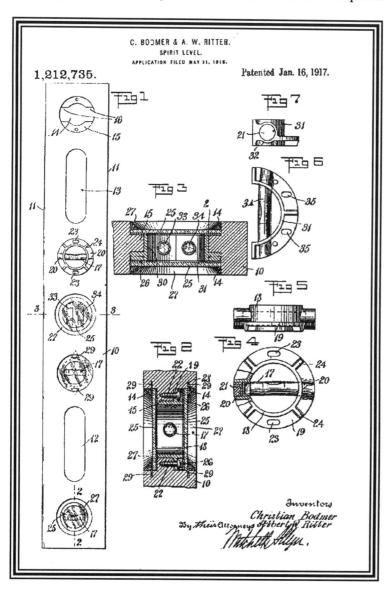

C. BODMER & A. W. RITTER.
SPIRIT LEVEL.
APPLICATION FILED MAY 31, 1916.

1,212,735.

Patented Jan. 16, 1917.

Inventors
Christian Bodmer
Albert W. Ritter
By their Attorneys

LEVEL

John Woods was listed as a farmer in the 1900 Census, and as a repairman with his own shop in the 1920 Census. He was 39 years old at the time the patent was issued.

This device is an attachment to turn a standard carpenters' level into a grade level. It specifies the addition of an adjustable foot to the level stock. The foot can be adjusted down using a captive nut on the shaft, accessible through a hole in the side of the level. The base of the shaft contains a crosshead with U-shape; this constitutes the foot. There seems to be nothing more to this device.

No example of this device has been identified to date. There are many devices of this type and most of them seem to be obscure.

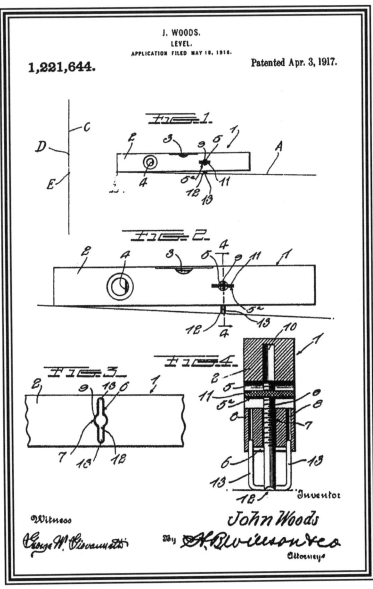

LEVEL

No listing for James Griffin was found in the New Britain City Directories during the period 1917-1920. However, in 1920 a James Griffin was found to be working for North & Judd Mfg. Co. (makers of saddlery hardware). He was working for Stanley by 1925.

The claims of the patent are that a stud *32* is inserted into the level stock and the vial case is placed over the stud and secured to the stock. Opposing screws bear against the stud and are used to bring the spirit vial into proper adjustment. A second claim of the patent is use of a the metal strip lengthwise over the spirit vial for protection.

This patent is for a simple adjustable vial case that could be attached to any straight edge. No example of this patent has been found at this time.

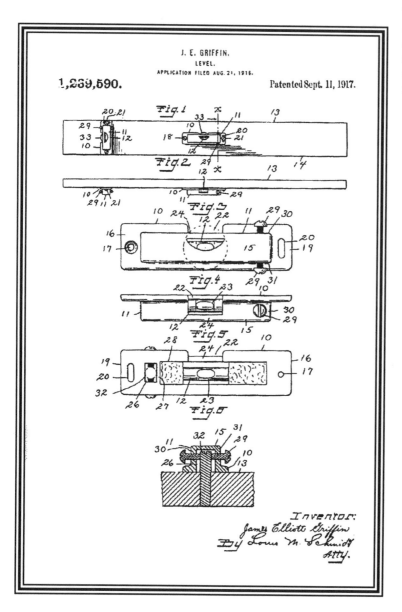

J. E. GRIFFIN,
LEVEL.
APPLICATION FILED AUG. 21, 1916.

1,239,590.

Patented Sept. 11, 1917.

Michael A. Costas
Biddeford, Maine

April 3, 1918
1,264,161

COMBINATION SQUARE, LEVEL, AND PROTRACTOR

Michael Costa [sic] was listed only in the 1920-21 Biddeford City Directory, and was gone by 1922-23. No city directory was available for the period 1913-19. He was a carpenter and maintained a listing in the business directory as well.

This patent seems to be built upon the Starrett patent No. 980,362 for a combined square and bevel protractor. In this case, one of the main additions is a spirit vial in a rotatable, graduated fixture in place of the two fixed vials in the Starrett patent. The frame is provided with a wedge screw, 25, for locking each fixture. The screws are turned with a captive nut, that is accessible through a cut out in the frame. It is, otherwise, a very similar device to that of Starrett, but it does seem to have a wider range of capabilities.

No example of this device has been identified to date.

PATENT DRAWINGS ON NEXT PAGE.

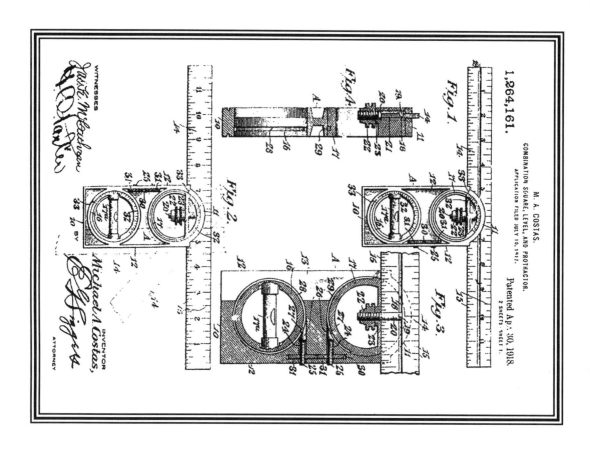

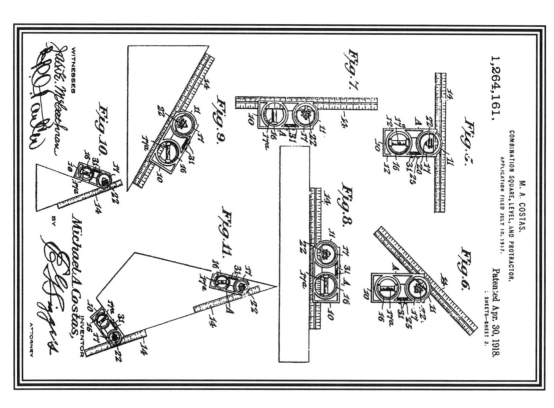

John P. Nikonow
Bridgeport, Connecticut

November 19, 1918
1,285,331

INCLINOMETER AND ACCELEROMETER

According to the patent papers, John Nikonow was a Russian citizen living in Bridgeport. However, he was never listed in the Bridgeport City Directories during the period 1914-1924.

The principle utilized in this patent requires two independent pendulums, one inverted, and a spring of proper tension. There are two separate scales, one fixed and one moving. One pendulum is affected by the incline of the vehicle and is read on the outer scale. The second pendulum is affected by the inertial changes due to acceleration or deceleration of the vehicle and the effect is read on the inner scale. *Fig. 5* shows a proposed arrangement of side-by-side rotating drums as an alternative.

The patent is for inclinometers and accelerometers for moving vehicles. It is not known if this device was ever manufactured, but it is an interesting application of inclinometers.

PATENT DRAWINGS ON NEXT PAGE.

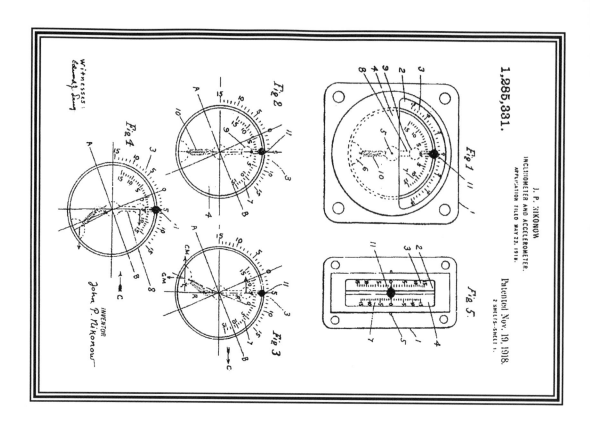

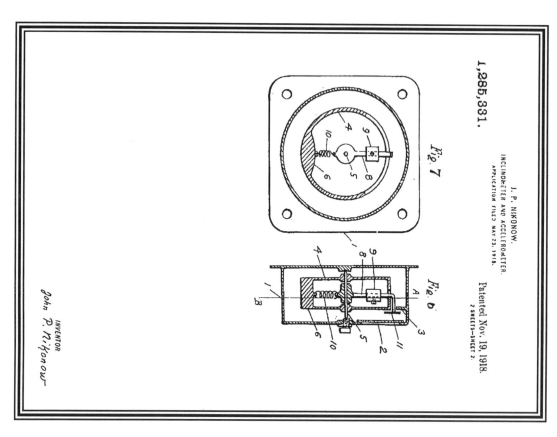

STOCK FOR SPIRIT LEVELS

It is believed that Gustav Vogel was a son (or brother) of William Vogel, and an older brother (or perhaps the father) of Paul Vogel. Beginning in 1906, he was shown as employed as a broom maker at William's home. He was listed with the same occupation through 1912. Listings continue through 1919, without an occupation. However, after 1920, he was recorded as a vice-president of P. H. Vogel Mfg. Co.

All of the claims for this patent involve a steel web plate and aluminum rails. The aluminum is to be cast over the edges of the steel plate. The steel web plate can be perforated or flanged, as shown in the drawings.

This patent, used on all Vogel levels, was used for the stock construction only. (Vogel levels were subsequently manufactured by Chapin-Stevens and finally by Stanley.) Vogel levels with steel webs are known but are much less common than those with brass webs. Levels carrying only this patent look exactly like the patent drawing, and are usually made with brass webs.

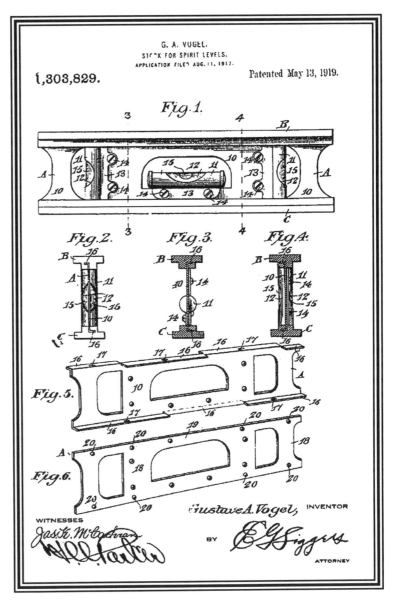

-209-

SPIRIT LEVEL AND GLASS

Napoleon Belleville was employed by the Lewiston Bleachery and Dye Works (which was actually in Lisbon, Maine). His trade was carpenter. From 1900 to 1904-05 he was just listed as a carpenter, independent of the Dye Works. He apparently started working at the Dye Works in 1908 (or earlier) and stayed there until his death in 1936. His history is remarkable, if only because in 10 listings for him in the Lewiston City Directories, he was found to have homes at 9 different addresses.

Belleville's patent specifies a manually setable inclinometer. A circular sleeve with graduations for 90° around its inner circumference is placed in a level stock. A cylinder carrying a spirit vial is fit into the sleeve. A pointer on the cylinder is arranged to indicate zero when the spirit vial is horizontally level. The pointer is also to serve as a handle for rotating the cylinder bearing the level.

The patent also specifies an interesting spirit vial (the "glass" in the title) whose ends gradually taper away from the center. The tapered ends wedge into collars on the cylinders and, thus, require no cement.

Although this device seems familiar, no examples are known at this time.

N. BELLEVILLE.
SPIRIT LEVEL AND GLASS.
APPLICATION FILED JAN. 18, 1918.

1,323,148.

Patented Nov. 25, 1919.

SPIRIT LEVEL AND PLUMB

For information about William Josep Neidl, see his patent of October 19 1909, on page 170.

The patent describes how a piece of sheet steel is rolled into a cylinder to contain the vial. The cylinder has a fin projecting downward for connecting to the stock. The vial cylinder is protected by two pieces of sheet steel that are formed around the sides of the cylinder and attached to the stock at the same point as the cylinder. (The Southington level - marked "patent applied for" - also utilizes a sheet steel stock and sheet steel pieces on each side of the vial cylinder that are fastened to the stock beneath the cylinder. The cylinder itself is not fastened to the stock.)

This patent covers vial holders and the method of mounting them to a stock. It is possible, even probable, that this is the basis for the Southington levels of similar construction. The Southington stock is, however, differently formed than implied by this patent. In spite of the implication gained by examining the patent drawings, no claim is made for construction of the stock.

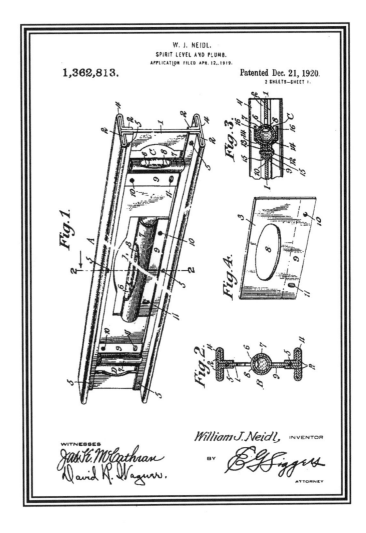

LEVEL-STOCK

For information about William Josep Neidl, see his patent of October 19, 1909 on page 170.

The patent concerns itself entirely with the construction of the stock, and the method of securing the sheet steel web to the sheet steel rails. It is primarily concerned with the two pairs of right angle plates that are to be welded to the web, and over which another plate is formed to produce the top and bottom rails.

To date, no example of the features of this patent have been identified.

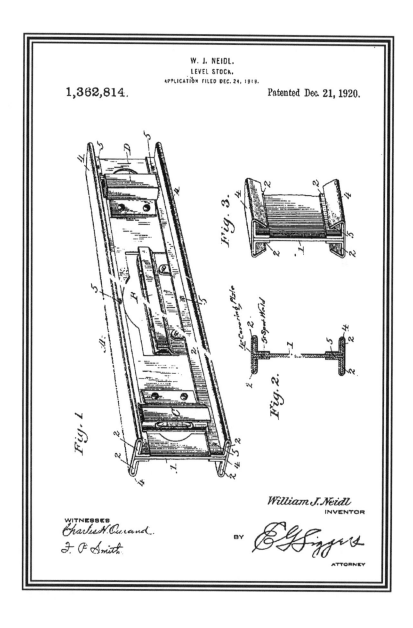

SPIRIT LEVEL

George Bedortha was the elder of the two Bedortha Brothers. He apparently supervised his father's (L. L. Bedortha) woodworking shop. It is now known that the Bedorthas produced levels, probably under contract for someone, as early as 1904.

The patent is for a level consisting of two light-weight rails (sheet metal channels) and two annular frames for vial holders. It is to be non-adjustable. The patent claims that it is important for rigidity that the circular frames be annular (cup-shaped).

No example of this level has been discovered to date. It would be a very cheap level as is emphasized in the patent.

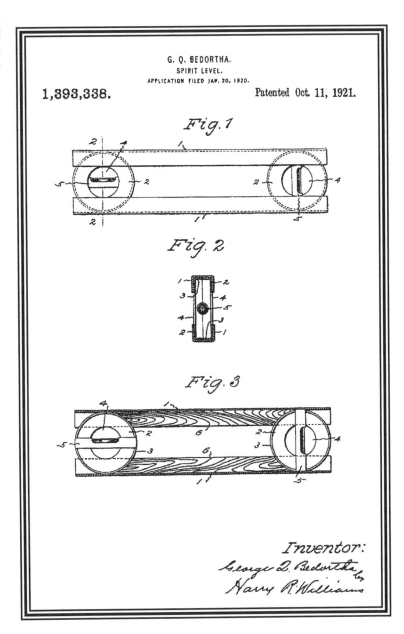

LEVEL ATTACHMENT

Tiffany was a very prominent businessman in Winsted. He was president of the Winsted Hosiery Co., Vice President of Hurlbut National Bank, and President and General Manager of New England Knitting Co. He died in 1928 at the age of 77.

This device in this patent is simply a captive nut on a screw that extends through the stock from top to bottom. A pointer is attached to the top of the screw at *12*, and a graduated scale is found on a side plate *7* above the nut. The intent of the device is to provide an adjustment feature that will enable the use of a carpenters' level as a grade level.

No specific example of this level attachment has been found to date.

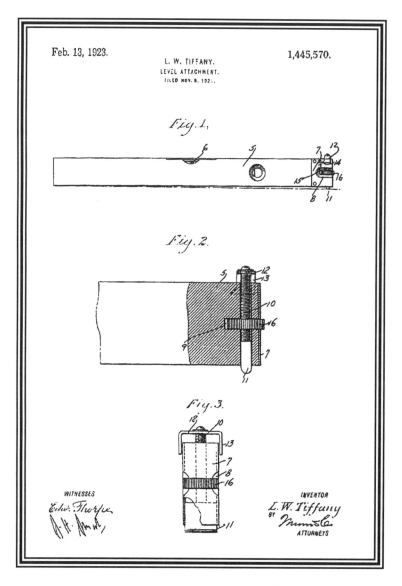

LEVEL ATTACHMENT

For information about Leverett W. Tiffany, see his patent of February 13, 1923, on page 214.

This patent is the second of three patents for similar devices granted to Mr. Tiffany. It could have been used as a grading attachment. The device is intended to replace one of the end plates. It consists of two pieces, and the inner piece is moved by hand. It can be locked in place with a knurled thumb screw. A scale is found on the outer piece, while the inner piece carries an indicator mark at *19*.

No example of this device has been found to date.

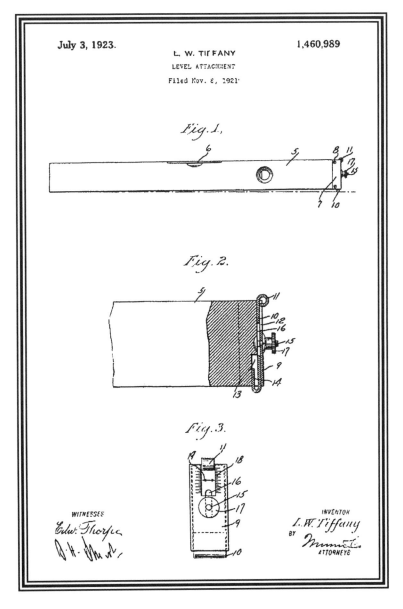

SPIRIT LEVEL

Paul Vogel at various times was an employee of P & F Corbin, and worked for Gustav Vogel as a broom maker. From 1915 to 1917, he was secretary-treasurer of New Britain Hardware, and in 1918 he was listed as an employee of New Britain Machine Company. He was not listed in the New Britain City Directories for 1919 or 1920. Paul is the son of William E. Vogel.

The patent describes vial fixtures that have the vial holder rigidly attached to the basic fitting, but the level can be adjusted by rotating the fitting. The plumb fittings are circular and mate with an empty circular piece on the opposite side of the level. The level fittings are roughly elliptical and have a vial carrier on each piece. The holes, through which the screws pass, are oversized, which allow adjustments to be made.

This was Paul Vogel's first patent and describes the vial holders used in Vogel Patent levels. Filed May 12, 1921, it took over two years for approval. All Vogel patent levels known to the author except those with brass webs carrying only the 1919 patent of Gustav Vogel, have vial holders identical to those proposed in this patent. Thus, only the steel web levels carrying the 1919 patent date have these vial holders.

Assigned to The Stanley Works of New Britain, CT

LEVEL STOCK

For information about Edmund A. Schade, see his patent of January 14, 1913 on page 173.

The patent describes a single piece metal stock. The stock is to have parallel straight edges, each reinforced by another and perpendicular flange. This second flange tapers down in width from the center towards the ends.

This patent is for a level stock and was used on the Stanley Nºs 233, 235, 236 and 313.

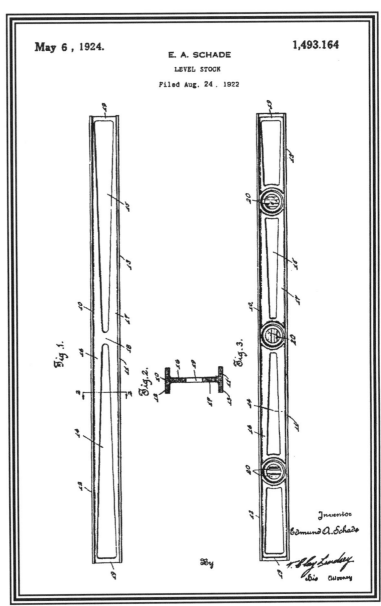

May 6 , 1924.

E. A. SCHADE

LEVEL STOCK

Filed Aug. 24 , 1922

1,493.164

Harris J. Cook
New Britain, Connecticut

November 11, 1924
1,515,239

Assigned to The Stanley Works of New Britain, CT

SPIRIT LEVEL

In 1924, Cook was a superintendent at Stanley. In 1931, he was just listed as a mechanical engineer in Stanley's employ.

This patent is for a simplified method of securing and adjusting a level vial. It is essentially a form of Traut's patent of May 8, 1906, except that instead of mounting the vial casket on springs to the bottom of the mortise in the stock, the vial casket was attached on springs to the top plate.

The patent was assigned to Stanley. This is another patent that required two years to move from initial submission to approval. It is not clear that this patent was ever utilized.

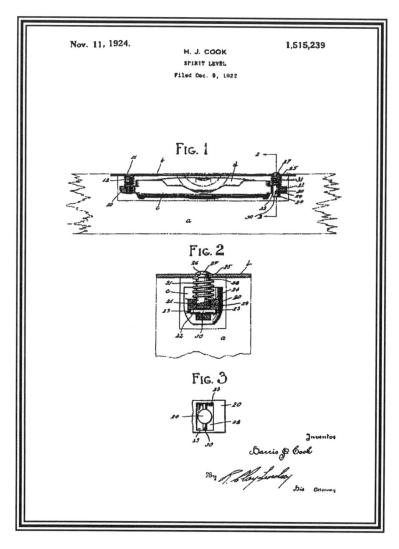

Robert A. Cecchini
Springfield, Massachusetts

May 26, 1925
1,539,543

Assigned one-quarter to George R. Cronin of Springfield, Mass.

PROTRACTOR LEVEL

Robert Cecchini was listed as a designer, in 1924, and as a mechanical engineer in 1925.

George R. Cronin was a pressman at Springfield Printing and Binding Co. (1920-1926)

The tool consists of a grooved base and a stock that carries a protractor. A fitting fastened at the pivot point of the protractor, and capable of rotating with the protractor, contains a receiver for a rod on the end of which, a spirit vial is attached. The protractor can rotate around a point at the center of a circle circumscribing the protractor segment. The level is carried at some distance from the center, which increases its sensitivity. An adjustment is provided on the right hand side for zeroing the device. A vernier scale is provided on the base stock of the tool.

The tool is envisioned with a micrometer attachment, as shown at *21*.

No example of this patent has been identified to date.

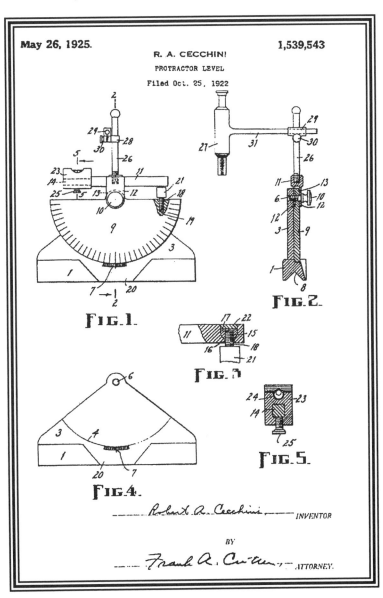

George Q. Bedortha
Windsor, Connecticut

December 1, 1925
1,563,322

SPIRIT-LEVEL GLASS

For information about George Q. Bedortha, see his patent of October 11, 1921 on page 213.

This patent is for use of a transparent, or yellow tinted, glass that absorbs "ultra violet and violet light rays." Such a glass, termed "Noviol" glass, was available at the time. An alternative calls for glass containing silver oxide. The objective is to prevent the early bleaching of the yellow fluorescent dyes (fluorescein or eosine) by ultraviolet light.

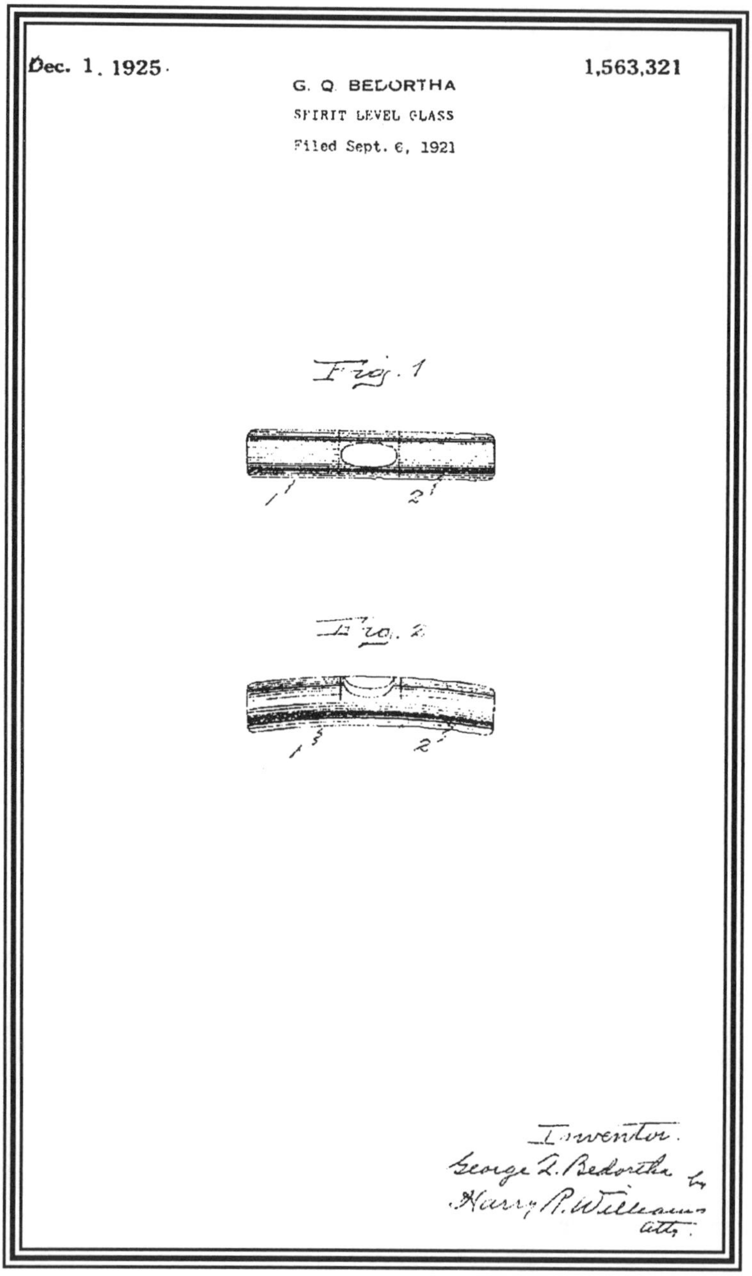

The patent application for this glass was filed on September 6, 1921 and took over four years to be approved. Certain Bedortha Brothers levels utilized a glass called Chromex, but it is not know if Chromex was in any way related to Noviol. Most spirit vials having yellow fluid at this time were still using clear glasses.

Dec. 1, 1925.

G. Q. BEDORTHA

SPIRIT LEVEL GLASS

Filed Sept. 6, 1921

1,563,321

Fig. 1

Fig. 2

Inventor.

George Q. Bedortha
by
Harry P. Williams
Atty.

Herman J. Halloran
Roslindale, Massachusetts

March 9, 1926
1,576,437

GRAVITY LEVEL

No information is available about Herman Halloran

This patent specifies a level with optics; however, the telescope is removably mounted, and the remainder of the instrument is still a functional inclinometer. The inclinometer is of the weighted wheel variety. The weighted wheel is graduated in degrees around its outer rim. This wheel is also connected by a shaft and two gears to another wheel graduated in minutes. The relationship of the gears is not specified, and the manner of marking the second wheel is left to the maker. Both wheels can be read through an opening in the top of the stock. The weight on the wheel is moveable so that it can be shifted if a different calibration is desired, or if the apparatus should become worn and require repositioning of the weight to remain in calibration.

No example of this device has been found to date.

PATENT DRAWINGS ON NEXT PAGE.

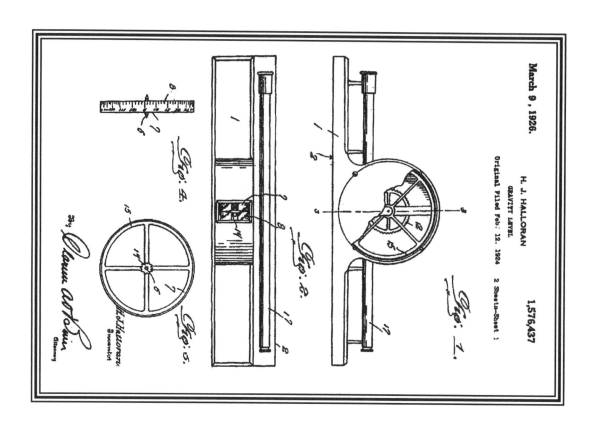

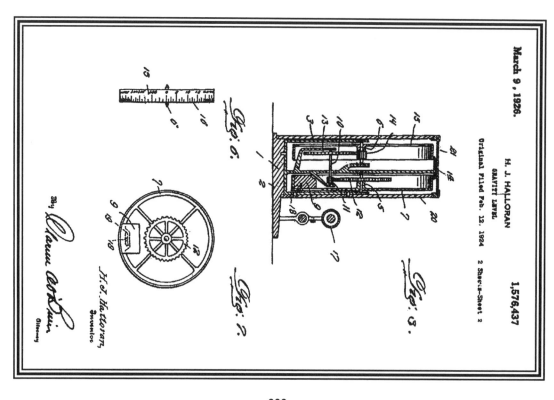

LEVEL ATTACHMENT

For information about Leverett W. Tiffany, see his patent of February 13, 1923, on page 214.

This grading attachment, unlike his previous patent, is two pieces with the base piece being stationary. Again unlike the previous patent, the scale is found on the base piece and is visible through a notch in the upper left corner of the moveable outer piece. The moveable part can be locked with a knurled thumb screw. This is a removeable device and need not become a permanent part of the level.

This is Tiffany's third patent and is also for a grading attachment. No examples of this attachment device have been encountered thus far.

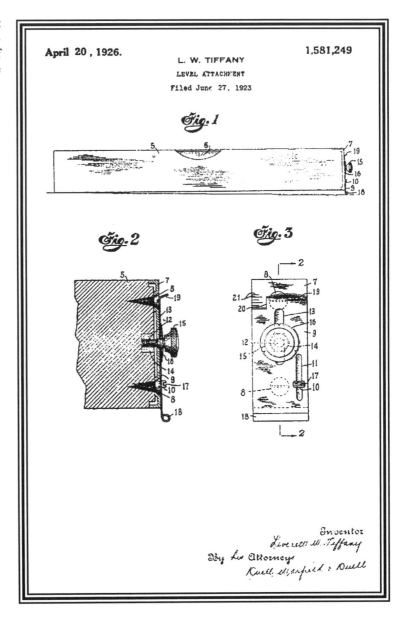

April 20, 1926.

L. W. TIFFANY

LEVEL ATTACHMENT

Filed June 27, 1923

1,581,249

Fig. 1

Fig. 2

Fig. 3

Inventor
Leverett W. Tiffany
By his Attorneys
Knell, Warfield & Duell

STOCK FOR SPIRIT LEVELS

For information on Paul Herman Vogel, see his patent of July 17, 1923 on page 216.

This patent covers the use of brass pieces for the web of the stock. The brass is to contain several small holes along its edges through which the aluminum for the rails may flow during casting, thus forming a bond.

The patent is for an improved stock, although the changes are not obvious. These stocks were used on Chapin-Stephens levels and, later, on Stanley levels. The patent application was filed October 10, 1923 and took over two and a half years for approval. It is presumed that Vogel, and later Chapin-Stephens, used the concept of this patent prior to approval being granted because Vogel levels with brass webs are well known.

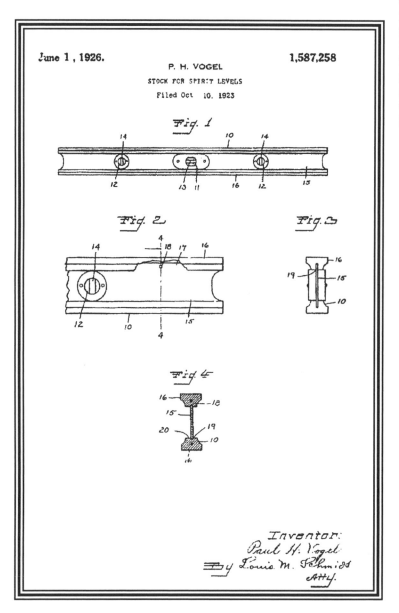

June 1, 1926.

P. H. VOGEL

STOCK FOR SPIRIT LEVELS

Filed Oct. 10, 1923

1,587,258

Fig. 1

Fig. 2

Fig. 3

Fig. 4

Inventor:
Paul H. Vogel
By Louis M. Schmidt
Atty.

John H. Carrier
New Britain, Connecticut

May 11, 1926
1,583,957

LEVEL

John Carrier was listed as a Stanley employee in the 1926 New Britain City Directory, but no other details of his life or employment are available. This patent, however, was not assigned to Stanley.

The patented invention consists of a spring placed under a vial holder (like a vial holder on a Stanley № 36) to aid in adjusting the vial. The claims of the patent call for a hole in the level frame, exactly like that used on the № 36, containing a tube held in place, and adjustable, by screws with cone shaped ends. In other words, the patent seems to be describing the same mechanism that Stanley had already been using for 30 years, with the addition of a flat spring under the vial cartridge. This patent required nearly two years for processing.

While this additional spring is unknown on Stanley levels, there are known to be other levels that use this exact type of flat spring under the vial cartridge. In addition, some rare Stanley № 36 levels are known that have a pivoting support under and in contact with the entire length of the vial cartridge. Such № 36 levels appear to be substantially later than the date of this patent.

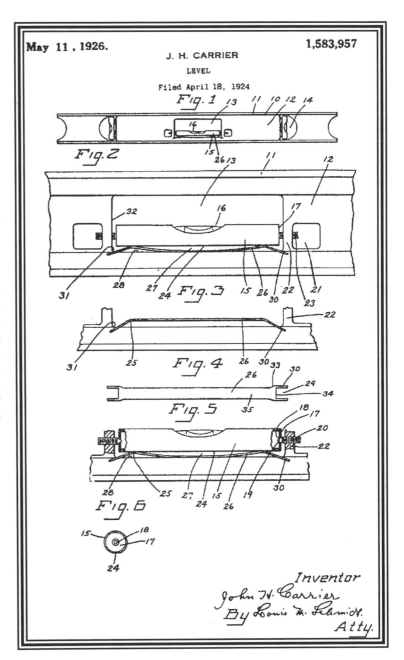

-225-

Oscar D Hapgood
Montague, Massachusetts

April 12, 1927
1,624,339

Assigned to the Goodell-Pratt Co. of Greenfield, MA

LEVEL

There is no information available regarding Oscar Hapgood.

The patent relates to a method of mounting a pair of spirit level vials in a receptacle, and to the adjustment of the vials in the carrier. Each fixture consists of two roughly semi-circular inner stocks, and two face plates. The inner stocks are pinned together on one end and forced apart by a spring on the other end. Each piece of the inner stock carries one spirit vial. The vials are independently adjustable, by turning a screw at *27*, when the fixture is assembled. The adjustment screws have conical points, and turning them has the effect of increasing or decreasing their wedging effect on the appropriate half of the inner stock. This, in turn, moves the spirit vial.

Goodell Pratt produced a level with this exact feature.

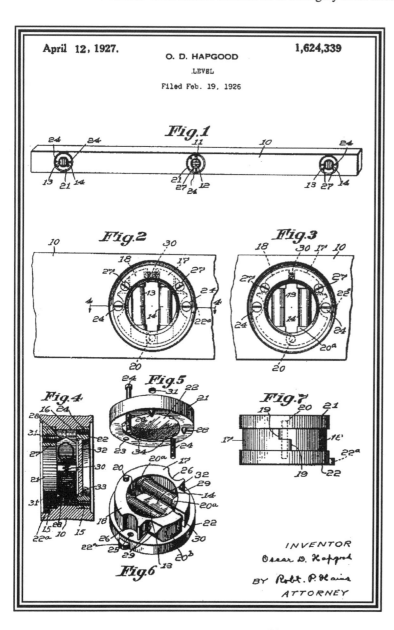

April 12, 1927.

O. D. HAPGOOD

.LEVEL

Filed Feb. 19, 1926

1,624,339

Fig.1

Fig.2

Fig.3

Fig.4

Fig.5

Fig.6

Fig.7

INVENTOR
Oscar D. Hapgood
BY Robt. P. Haire
ATTORNEY

Paul Herman Vogel
New Hartford, Connecticut

October 30, 1928
1,689,982

Assigned to the Chapin Stephens Co. of Pine Meadow, CT

LEVEL VIAL HOLDER

For information on Paul Herman Vogel, see his patent of July 17, 1923 on page 216.

The patent is for an improved level vial holder for wooden levels. In intent, this patent is similar to the 1917 patent of Bodmer and Ritter. The level stock is bored through, and then mortised around the edges of that hole. A plate bearing a spirit vial is inserted into the hole from each side and fastened there with machine screws. The plates contain slightly oversized holes for the bolts, thus allowing for adjustment. A cover plate is attached over each side with screws.

This Vogel patent was for improved vial holders. It was applied for and obtained after the P. H. Vogel Co. was acquired by Chapin-Stephens. No levels marked with this patent date have been identified.

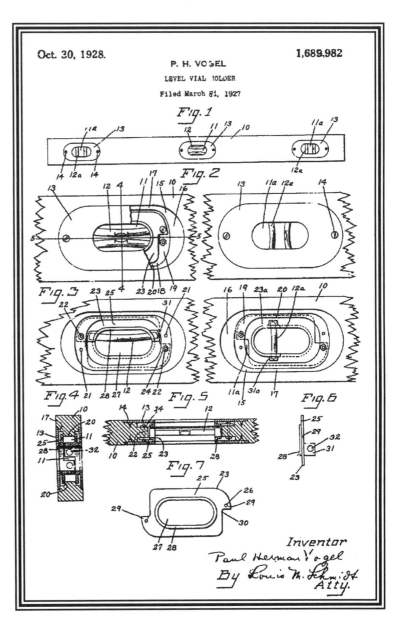

-227-

Austin L. Stowell
New Britain, Connecticut

June 11, 1929
1,716,744

Assigned to The Stanley Works of New Britain, CT

LINE LEVEL

Stowell worked as a mechanical engineer at Stanley Rule & Level. There is little else of note about Mr Stowell. Note that it took almost three years for this patent to be approved.

The patent describes an aluminum line level with a U-shaped hook at each end and openings opposing each other on each leg of the U. This is best shown in *Fig.5*. A second feature of the patent was the plugs to which the U-shaped hooks were attached, as shown in *Fig.5*. There are seven claims made in this patent, all of them regarding one or the other of these two features.

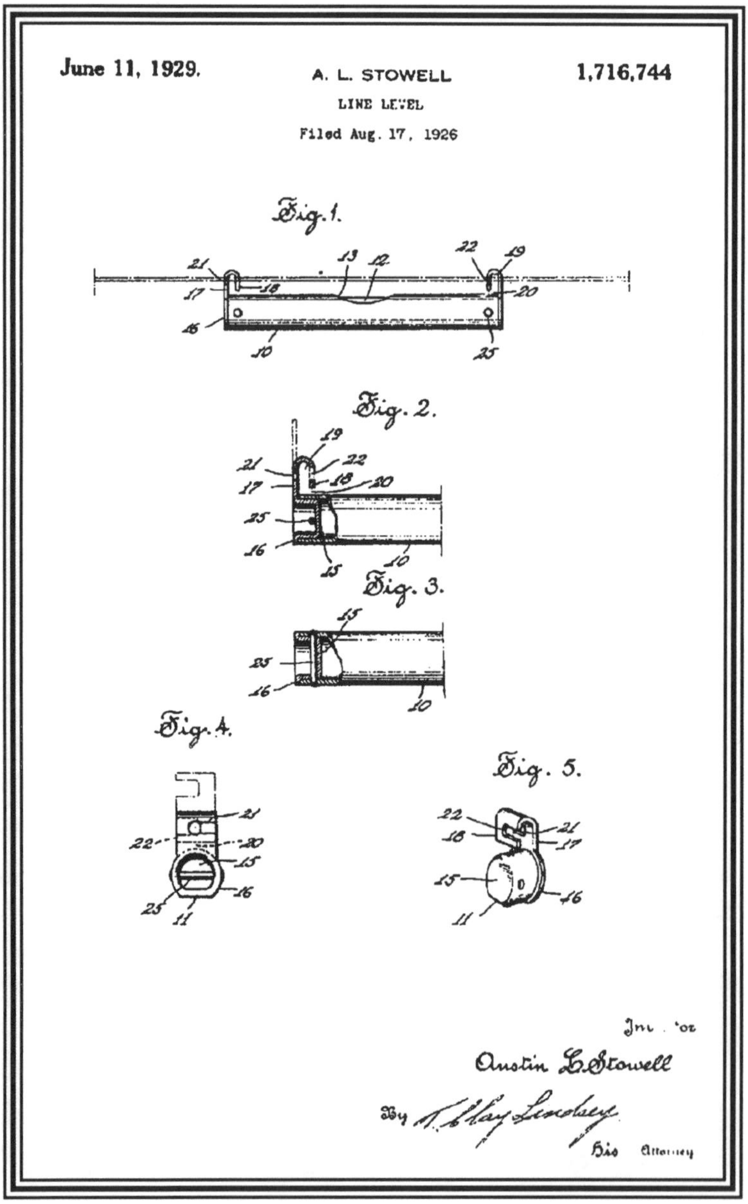

Stanley sold a line level based on this patent as Nº 187. The design was superseded by the patent of Frederick Volz (September 17, 1935) and subsequent Nº 187s were made according to the Volz patent.

June 11, 1929.

A. L. STOWELL
LINE LEVEL
Filed Aug. 17, 1926

1,716,744

Fig. 1.

Fig. 2.

Fig. 3.

Fig. 4.

Fig. 5.

Austin L. Stowell
By T. Clay Lindsey
His Attorney

Charles D. Severance and Raymond O. Stetson
Greenfield, Massachusetts

March 25, 1930
1,752,112

Assigned to the Goodell-Pratt Co. of Greenfield, Mass

SPIRIT LEVEL

Charles Severance was a long time employee of Goodell Pratt. He held the office of Secretary until 1912, after which he is shown as a Vice-President. He had previously been a bookkeeper at Wells Bros. and Company.

Raymond O. Stetson was discussed in regard to his patent of July 18, 1905 on page 161.

This patent, while presenting a method of mounting a pair of spirit level vials in a receptacle, and the receptacle in a level stock, also presents a means of adjusting the point at which level is read. The level vials pass through the walls of the receptacle and are anchored into the stock. The adjustment consists of a moveable bead on a wire, much as the early Stratton Bros. patent required. Two beads per wire may also be used to frame the bubble. The patent also specifies that there should be glass windows over the vials, and is the first of the New England patents to make such a specification.

This patent was incorporated into the 6000 Series of Goodell Pratt six-glass levels.

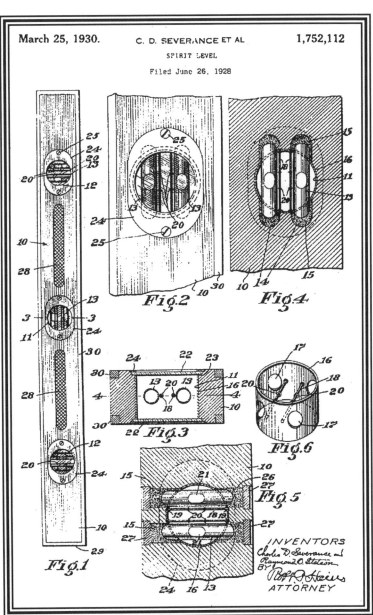

March 25, 1930.   C. D. SEVERANCE ET AL   1,752,112
SPIRIT LEVEL
Filed June 26, 1928

Fig.1 *Fig.2* *Fig.3* *Fig.4* *Fig.5* *Fig.6*

INVENTORS
Charles D. Severance and
Raymond O. Stetson
BY
ATTORNEY

Louis B. Beecher and Patrick Ahern
New Britain, Connecticut

September 15, 1931
1,823,524

Assigned to The Stanley Works of New Britain, CT

SPIRIT LEVEL

In 1924, Louis Beecher was a Superintendent for Stanley in Kensington, CT.

Patrick Ahern was a foreman for Stanley in New Britain in 1924, and in 1931 he was made a superintendent.

This patent describes a means for providing fixed vials in a wooden level stock. The vials are held in position by the ferrules shown in *Figs.6* and *7,* and the whole assembly is cemented into the level stock. A glass plate is held in place over each side of the ferrule by means that are not described.

This patent application was filed March 16, 1926, meaning that it required 5½ years to win approval. The patent was used on the 250- series of masons' levels and concerned a modern vial assembly. Stanley began to use the device as soon as the patent was filed.

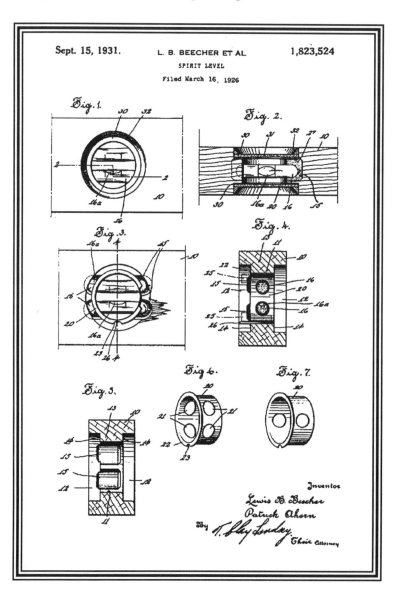

Sept. 15, 1931.   L. B. BEECHER ET AL   1,823,524

SPIRIT LEVEL

Filed March 16, 1926

Harry P. Swain
Concord, New Hampshire

September 27, 1932
1,879,587

LEVEL AND PLUMB INDICATOR

Harry P. Swain first appeared in the Concord City Directories in 1906-07. He did not appear in the 1932 directory. He appeared to be a mason as was his father, Stephen.

The invention (an inclinometer) comprises a two-sided (two dials) chamber containing a graduated dial for viewing from each side. The only thing showing is one hand (like a clock hand) on each face. The center of the device contains a weight suspended from the same shaft as the hands. The weight bears an epicentric gear which meshes with a gear on one end of the shaft, and with teeth on the inside of the housing. The gearing serves to multiply the movement of the plumb weight by an amount proportional to the number of teeth of the various gears (not specified). The movements of the pointer are thus exaggerated, and the dial may be calibrated in units (not specified) that are easier and more accurate to read.

No example of this device has been identified to date. The patent language is exceeding difficult to understand, and it took over two years for it to be granted.

PATENT DRAWINGS ON NEXT PAGE.

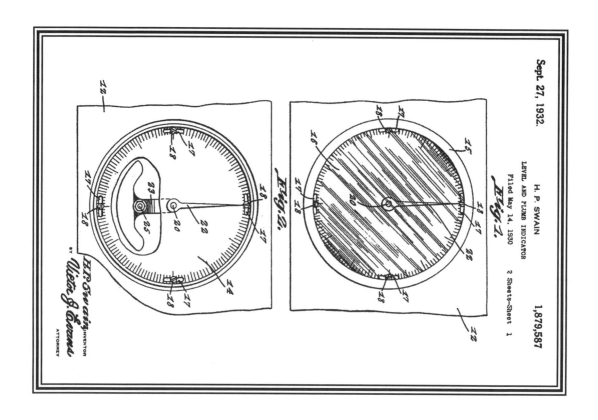

Sept. 27, 1932.

H. P. SWAIN

LEVEL AND PLUMB INDICATOR

Filed May 14, 1930

2 Sheets-Sheet 1

1,879,587

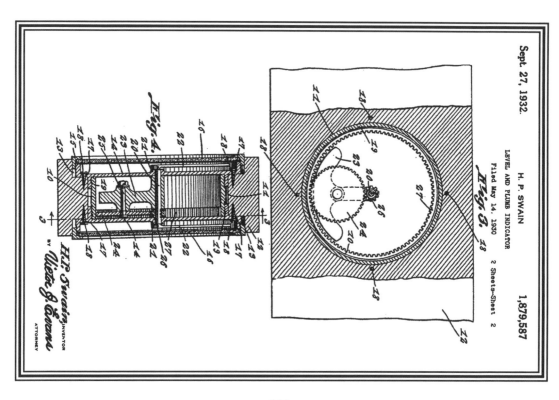

Sept. 27, 1932.

H. P. SWAIN

LEVEL AND PLUMB INDICATOR

Filed May 14, 1930

2 Sheets-Sheet 2

1,879,587

ALPHABETICAL INDEX BY PATENTEE

Patentee	Patent Date	Patent Nº	Page
Franchini, Alcide	06-14-1910	961599	175
Gallis, Rudolph Peter	02-13-1883	272231	53
Goodell, Albert D.	12-30-1884	310046	67
Goodell, Albert D.	10-16-1888	391242	95
Green, Henry	10-14-1890	438541	104
Green, Henry	03-31-1891	449609	105
Green, Henry	04-14-1891	450457	106
Griffen, James Elliott	09-11-1917	1239590	204
Hall, George F. & Traut, Justus A.	09-18-1888	389647	94
Halloran, Herman J.	03-09-1926	1576437	221-222
Hamalainen, Matti	06-08-1915	1142419	192
Hapgood, Oscar D.	04-12-1927	1624339	226
Harmon, John W.	11-23-1880	234709	50
Harmon, John W.	01-23-1883	270951	52
Harmon, John W.	07-17-1883	281267	54-55
Holbrook, Sylvanus	10-11-1875	168567	39
Humphrey, Ira	05-27-1879	215749	48
Hutton, James C.	07-03-1888	385516	92-93
Hyde, Albe F.	09-01-1874	154677	37
Janssen, C. Diedrich	11-17-1885	330589	73
Janssen, Carsten Diedrich	12-17-1901	688952	145
Johnson, Alexander	05-02-1911	991446	181
Kelly, Harold	11-19-1901	686975	144
Ladd, William G. Jr	04-09-1850	7263	1
Lawrence, Daniel F.	02-28-1911	985685	180
Lewis, Homer	06-30-1868	79363	24
Littlefield, J. A.	04-05-1870	101477	31
Long, Charles Benjamin	11-08-1887	372921	88
Loomis, Hiram G.	07-07-1868	79582	25
Lord, Peter	06-07-1904	762062	158
Lowe, Harry	11-05-1889	414232	98
Lyons, George W.	11-21-1916	1205947	200
McDowell, Albert F.	12-12-1905	806987	162
McLaughlin, Joseph	01-21-1890	419703	99
Makowski, Albert	02-27-1912	1018719	184
Martin, William	03-18-1890	423484	101
Merrill, Bisbee B.	02-24-1885	312743	68
Merrill, Nathaniel C.	03-21-1899	621358	136
Miller, Frank B.	11-09-1915	1159522	195
Morton, William M.	12-22-1896	573682	126
Morton, William M.	02-23-1897	577772	127
Morton, William M.	05-04-1897	581938	128
Morton, William M.	10-05-1897	591139	130
Neidl, William J.	10-19-1909	937633	170-171
Neidl, William J.	03-28-1916	1177131	196
Neidl, William Josep	12-21-1920	1362813	211
Neidl, William Josep	12-21-1920	1362814	212
Nicholson, William T.	05-01-1860	28104	5
Nikonow, John P.	11-19-1918	1285331	207-208
Odholm, A.P.	10-26-1869	96140	29
Oliver, Moses F.	01-12-1892	467016	108
Parkhurst, Charles J. & Parkhurst, Albert W.	04-08-1884	296608	61
Pickering, Oliver	12-18-1877	198311	46
Poole, Nathaniel D.	12-06-1892	487427	111
Potter, Charles Mortimer	12-28-1897	596279	133
Praddex, William	02-23-1892	469451	109

Patentee	Patent Date	Patent №	Page
Quinby, George F.	09-11-1900	657675	138
Raymond, Lewis C.	08-21-1895	544321	118
Reilly, Dennis E.	10-04-1910	971792	177-178
Rhoades, Alonzo E.	06-30-1903	732223	151
Rich, Henry Murdock	10-14-1884	306429	66
Richardson, Charles F.	08-16-1887	369308	82
Richardson, William L.	01-01-1867	60788	11
Ritchie, Frederick W.	08-19-1884	303666	65
Root, Albert	04-29-1884	297719	63-64
Sanderson, Samuel	10-21-1873	143993	36
Sanford, L. Arthur	03-09-1886	337621	74
Sawyer, Burnside E.	12-27-1898	29875	135
Schade, E.A.	09-22-1914	1111677	189
Schade, E.A.	08-24-1915	1151549	193-194
Schade, Edmund A.	01-09-1917	1211882	201
Schade, Edmund	05-06-1924	1493164	217
Scott, John & McGrath, Nicholas B.	05-13-1913	1061638	186
Severance, Charles D. & Stetson, Raymond O.	01-25-1930	1752112	229
Seymour, Daniel J.	06-13-1911	995099	182
Shelley, George A.	11-07-1871	120675	32
Sheflott, Leonard C.	11-13-1906	835986	165
Shepardson, H. S.	09-13-1864	44225	9
Sibley, J.D.	06-23-1868	79226	23
Starrett, Laroy S.	05-06-1879	215024	47
Starrett, Laroy S.	08-07-1883	282583	56
Starrett, Laroy S.	04-14-1896	25392	120
Starrett, Laroy S.	11-08-1898	613946	134
Starrett, Laroy S.	12-27-1904	778808	159
Starrett, Laroy S.	01-03-1911	980362	179
Starrett, Laroy S.	10-24-1916	1202114	199
Stephens, L.C.	01-12-1858	19105	4
Stetson, Raymond O.	07-18-1905	794753	161
Stevens, Fred	02-07-1905	781749	160
Stowell, Austin L.	06-11-1929	1716744	228
Stratton Edwin A. & Charles M.	03-01-1870	100463	30
Stratton Edwin A. & Charles M.	07-16-1872	129183	35
Stratton, Edwin A. & Charles, M.	10-04-1887	370826	86
Stratton, Edwin A. & Charles, M.	05-22-1888	383196	91
Swain, Harry P.	09-27-1932	1879587	231-232
Tate, William J.	01-21-1868	73670	20
Taylor, Augustus G.	05-03-1892	474152	110
Thayer, Eli	08-26-1862	36312	7
Thibert, Napoleon R.	12-07-1909	942114	172
Tiffany, Leverett W.	02-13-1923	1445570	214
Tiffany, Leverett W.	07-03-1923	1460989	215
Tiffany, Leverett W.	04-20-1926	1581249	223
Tiller, Charles	08-16-1887	368434	83-84
Tilley, Thomas Seymour	10-21-1902	711801	147-148
Tilley, Thomas Seymour	10-06-1903	740491	153-154
Traut, Frederick A.	04-21-1863	38252	8
Traut, Justus A.	10-06-1868	82769	28
Traut, Justus A.	07-02-1872	128513	34
Traut, Justus A.	11-16-1886	352721	77
Traut, Justus A.	05-07-1889	402869	96
Traut, Justus A.	06-18-1889	405624	97
Traut, Justus A.	02-18-1890	421786	100

Patentee	Patent Date	Patent №	Page
Traut, Justus A.	03-25-1890	423969	102
Traut, Justus A.	06-02-1891	453452	107
Traut, Justus A.	12-26-1893	511377	112
Traut, Justus A.	07-17-1894	523021	113
Traut, Justus A.	07-17-1894	523022	114
Traut, Justus A.	07-17-1894	523023	115
Traut, Justus A.	02-19-1895	534303	116
Traut, Justus A.	02-04-1896	25130	119
Traut, Justus A.	08-04-1896	565096	123
Traut, Justus A.	08-04-1896	565097	124
Traut, Justus A.	08-04-1896	565098	125
Traut, Justus A.	05-14-1901	674107	141
Traut, Justus A.	07-01-1902	703678	146
Traut, Justus A.	01-27-1903	719061	149
Traut, Justus A.	04-28-1903	726377	150
Traut, Justus A.	05-24-1904	760587	156
Traut, Justus A.	05-08-1906	820028	163
Traut, Justus A. & Bodmer, Christian	06-23-1896	562678	121
Traut, Justus A. & Bodmer, Christian	06-23-1896	562679	122
Traut, Justus A. and Frank L.	11-03-1908	902778	167
Van Alstyne, Lawrence	10-23-1883	287342	58
Vogel, Gustav A.	05-13-1919	1303829	209
Vogel, Paul Herman	07-17-1923	1462430	216
Vogel, Paul H.	06-01-1926	1587259	224
Vogel, Paul Herman	10-30-1928	1689982	227
Vose, Ambrose S.	07-06-1886	345196	76
Walsh, Jas.; Murphy, Tho. F.; & Clark, Everett A.	05-05-1885	317250	70-71
Webb, Edward E.	12-07-1886	354076	78-79
Wickham, Almeron W. & Roach, James M.	11-22-1887	373627	89-90
Wilbar, Francis	04-20-1852	8897	2
Wilcox, Martin	06-06-1876	178354	40
Winter, Ignatius S.	11-04-1879	221380	49
Woods, George	10-11-1887	371294	87
Woods, John	04-03-1917	1221644	203
Woolson, Harry H.	01-05-1909	909046	168-169

CHRONOLOGICAL INDEX

Patent Date	Patent Nº	Patentee	Page
12-30-1884	310046	Goodell, Albert D.	67
02-24-1885	312743	Merrill, Bisbee B.	68
04-07-1885	315264	Euington, Matthew	69
05-05-1885	317250	Walsh, James; Murphy, Thomas F.; & Clark, Everett A.	70-71
10-27-1885	329156	Ford, William	72
11-17-1885	330589	Janssen, C. Diedrich	73
03-09-1886	337621	Sanford, L. Arthur	74
04-06-1886	339158	Finley, John A.	75
07-06-1886	345196	Vose, Ambrose S.	76
11-16-1886	352721	Traut, Justus A.	77
12-07-1886	354076	Webb, Edward E.	78-79
12-21-1886	354592	Brown, William Wood	80
03-29-1887	360378	Duffy, Edward	81
08-16-1887	368308	Richardson, Charles F.	82
08-16-1887	368434	Tiller, Charles	83-84
09-20-1887	370054	Fox, Philo L.	85
10-04-1887	370826	Stratton, Edwin A. & Stratton, Charles, M.	86
10-11-1887	371294	Woods, George	87
11-08-1887	372921	Long, Charles Benjamin	88
11-22-1887	373627	Wickham, Almeron W. & Roach, James M.	89-90
05-22-1888	383196	Stratton, Edwin A. & Stratton, Charles, M.	91
07-03-1888	385516	Hutton, James C.	92-93
09-18-1888	389647	Hall, George F. & Traut, Justus A.	94
10-16-1888	391242	Goodell, Albert D.	95
05-07-1889	402869	Traut, Justus A.	96
06-18-1889	405624	Traut, Justus A.	97
11-05-1889	414232	Lowe, Harry	98
01-21-1890	419703	McLaughlin, Joseph	99
02-18-1890	421786	Traut, Justus A.	100
03-18-1890	423484	Martin, William	101
03-25-1890	423969	Traut, Justus A.	102
09-16-1890	436495	Fox, Philo L.	103
10-14-1890	438541	Green, Henry	104
03-31-1891	449609	Green, Henry	105
04-14-1891	450457	Green, Henry	106
06-02-1891	453452	Traut, Justus A.	107
01-12-1892	467016	Oliver, Moses F.	108
02-23-1892	469451	Praddex, William	109
05-03-1892	474152	Taylor, Augustus G.	110
12-06-1892	487427	Poole, Nathaniel D.	111
12-26-1893	511377	Traut, Justus A.	112
07-17-1894	523021	Traut, Justus A.	113
07-17-1894	523022	Traut, Justus A.	114
07-17-1894	523023	Traut, Justus A.	115
02-19-1895	534303	Traut, Justus A.	116
06-18-1895	541151	Carlson, Carl A.	117
08-21-1895	544321	Raymond, Lewis C.	118
02-04-1896	25130	Traut, Justus A.	119
04-14-1896	25392	Starrett, Laroy S.	120
06-23-1896	562678	Traut, Justus A. & Bodmer, Christian	121
06-23-1896	562679	Traut, Justus A. & Bodmer, Christian	122
08-04-1896	565096	Traut, Justus A.	123
08-04-1896	565097	Traut, Justus A.	124
08-04-1896	565098	Traut, Justus A.	125
12-22-1896	573682	Morton, William M.	126
02-23-1897	577772	Morton, William M.	127
05-04-1897	581938	Morton, William M.	128
05-11-1897	582517	Bellows, Stephen H.	129
10-05-1897	591139	Morton, William M.	130
10-05-1897	591153	Berger, Christian L.	131
10-26-1897	592537	Carriere, Joseph	132
12-28-1897	596279	Potter, Charles Mortimer	133
11-08-1898	613946	Starrett, Laroy S.	134

Patent Date	Patent №	Patentee	Page
12-27-1898	29875	Sawyer, Burnside E.	135
03-21-1899	621358	Merrill, Nathaniel C.	136
06-19-1900	652078	Cable, Frank T.	137
09-11-1900	657675	Quinby, George F.	138
12-04-1900	663252	Bogardus, John S.	139-140
05-14-1901	674107	Traut, Justus A.	141
09-24-1901	683254	Delin, Andre	142
10-29-1901	685569	Bullard, James H.	143
11-19-1901	686975	Kelly, Harold	144
12-17-1901	688952	Janssen, Carsten Diedrich	145
07-01-1902	703678	Traut, Justus A.	146
10-21-1902	711801	Tilley, Thomas Seymour	147-148
01-27-1903	719061	Traut, Justus A.	149
04-28-1903	726377	Traut, Justus A.	150
06-30-1903	732223	Rhoades, Alonzo E.	151
09-29-1903	740255	Dillon, Herbert T.	152
10-06-1903	740491	Tilley, Thomas Seymour	153-154
10-06-1903	740742	Bush, James F.	155
05-24-1904	760587	Traut, Justus A.	156
05-24-1904	761033	Cross, Anson K.	157
06-07-1904	762062	Lord, Peter	158
12-27-1904	778808	Starrett, Laroy S.	159
02-07-1905	781749	Stevens, Fred	160
07-18-1905	794753	Stetson, Raymond O.	161
12-12-1905	806987	McDowell, Albert F.	162
05-08-1906	820028	Traut, Justus A.	163
11-06-1906	834964	Broderick, David Felix	164
11-13-1906	835986	Sheflott, Leonard C.	165
06-30-1908	892217	Brown, Hayden William	166
11-03-1908	902778	Traut, Justus A. and Frank L.	167
01-05-1909	909046	Woolson, Harry H.	168-169
10-19-1909	937633	Neidl, William J.	170-171
12-07-1909	942114	Thibert, Napoleon R.	172
04-05-1910	954074	Bodmer, Christian & Burdick, James M.	173
05-24-1910	959296	Bodmer, Christian	174
06-14-1910	961599	Franchini, Alcide	175
09-20-1910	970472	DeBisschop, Joseph E.	176
10-04-1910	971792	Reilly, Dennis E.	177-178
01-03-1911	980362	Starrett, Laroy S.	179
02-28-1911	985685	Lawrence, Daniel F.	180
05-02-1911	991446	Johnson, Alexander	181
06-13-1911	995099	Seymour, Daniel J.	182
12-12-1911	1011262	Starrett, Laroy S.	183
02-27-1912	1018719	Makowski, Albert	184
01-14-1913	1050610	Burdick, James M. & Schade, Edmund	185
05-13-1913	1061638	Scott, John & McGrath, Nicholas B.	186
06-03-1913	1063342	Ekman, Per Erik	187-188
09-22-1914	1111677	Schade, E. A.	189
01-12-1915	1124833	Bartlett, George W.	190-191
06-08-1915	1142419	Hamalainen, Matti	192
08-24-1915	1151549	Schade, E. A.	193-194
11-09-1915	1159522	Miller, Frank B.	195
03-28-1916	1177131	Neidl, William J.	196
06-06-1916	1186064	Arkin., Louis	197
07-04-1916	1189422	Bodmer, Christian A.	198
10-24-1916	1202114	Starrett, Laroy S.	199
11-21-1916	1205947	Lyons, George W.	200
01-09-1917	1211882	Schade, Edmund A.	201
01-16-1917	1212735	Bodmer, Christian A. & Ritter, Albert W.	202
04-03-1917	1221644	Woods, John	203
09-11-1917	1239590	Griffin, James Elliott	204
04-30-1918	1264161	Costas, Michael A.	205-206
11-19-1918	1285331	Nikonow, John P.	207-208
05-13-1919	1303829	Vogel, Gustav A.	209
11-25-1919	1323148	Belleville, Napoleon	210

Index by Category

+ = additional patentee(s)

Inclinometers

Design Patents

Patentee	Patent Date	Page

Construction Technique, Detail or Feature

William T. Nicholson	May 1, 1860	5
Edwin A. Stratton	March 1, 1870	30
Albe F. Hyde	Sept. 1, 1874	37
Edwin L. Barnes	Feb. 16, 1875	38
Burkner F. Burlington +	Oct. 3, 1876	41
Nelson H. Bearse	Sept. 11, 1883	57
Edward E. Webb	Dec. 7, 1886	78
Philo L. Fox	Sept. 20, 1887	83
Edwin A. Stratton +	Oct. 4, 1887	86
Charles Benjamin Long	Nov. 8, 1887	88
Philo L. Fox	Sept. 16, 1890	103
Justus A. Traut	June 2, 1891	107
William Praddex	Feb. 23, 1892	109
Justus A. Traut	Dec. 26, 1893	112
Justus A. Traut	July 17, 1894	113
Justus A. Traut	July 17, 1894	114
Justus A. Traut	July 17, 1894	115
Justus A. Traut	Feb. 19, 1895	116
Justus A. Traut	Aug. 4, 1896	124
Justus A. Traut	Aug. 4, 1896	125
William M. Morton	Oct. 5, 1897	130
Justus A. Traut	July 1, 1902	146
Justus A. Traut	May 24, 1904	156
Justus A. Traut +	Nov. 3, 1908	167
Christian Bodmer +	April 5, 1910	173
Christian Bodmer	May 24, 1910	174
Daniel J. Seymour	June 13, 1911	182
Edmund A. Schade	Sept. 22, 1914	189
Frank B. Miller	Nov. 9, 1915	195
William Josep Neidl	March 28, 1916	196
George W. Lyons	Nov. 21, 1916	200
Christian Bodmer +	Jan. 16, 1917	202
James Elliott Griffin	Sept. 11, 1917	204
William Josep Neidl	Dec. 21, 1920	211
"	"	212
George Q. Bedortha	Oct. 11, 1921	213
Paul Herman Vogel	July 17, 1923	216
Edmund Schade	May 6, 1924	217
Paul Herman Vogel	June 1, 1926	224
Oscar D. Hapgood	April 12, 1927	226
Paul Herman Vogel	Oct. 30, 1928	227
Charles D. Severance +	March 25, 1930	229
Louis B. Beecher +	Sept. 15, 1931	230

Grade Levels or Attachments

Lebbius Brooks	Aug. 20, 1854	3
Alvin Cahoon, Jr.	May 20, 1862	6
Sylvanus Holbrook	Oct. 11, 1875	39
Stephen H. Bellows	May 11, 1897	128
Charles M. Potter	Dec. 28, 1897	133
James Bullard	Oct. 29, 1901	143
Justus H. Traut	Jan. 27, 1903	149
Herbert T. Dillon	Sept. 29, 1903	152
Albert F. McDowell	Dec. 12, 1905	162
David Felix Broderick	Nov. 6, 1909	164
Joseph E. DeBisschop	Sept. 20, 1910	176
John Woods	April 3, 1917	203
Leverett W. Tiffany	Feb. 13, 1923	214
Leverett W. Tiffany	July 3, 1923	215
Leverett W. Tiffany	April 20, 1926	223

Patentee	Patent Date	Page

Plumb Levels

Charles Tiller	Aug. 16, 1887	84
Lewis C. Raymond	Aug. 13, 1895	118
Carsten Diedrich Janssen	Dec. 17, 1901	145
Alexander Johnson	May 2, 1911	181

Sighting Levels, Features or Attachments

Aaron Chase, Jr.	August 29, 1861	10
William L. Richardson	Jan. 1, 1867	11
J.D. Sibley	June 23, 1868	23
Hiram G. Lewis	July 17, 1868	25
Oliver Pickering	Dec. 18, 1877	46
John W. Harmon	Nov. 23, 1880	50
John W. Harmon	Jan. 23, 1883	52
Rudolph Peter Gallis	Feb. 13, 1883	53
John W. Harmon	July 17, 1883	54
Charles F. Richardson	Aug. 16, 1887	82
Almeron W. Wickham	Nov. 22, 1887	89
Justus A. Traut	May 7, 1889	96
Justus A. Traut	June 18, 1889	97
Hayden William Brown	June 30, 1908	166
Edmund A. Schade	Aug. 24, 1915	193
Christian Bodmer	July 4, 1916	198
Herman J. Halloran	March 9, 1926	221

Adjustment Features

Samuel S. Chapin +	Sept. 10, 1867	14
William Tate	Jan. 21, 1868	20
L.L. Davis	March 17, 1868	21
L.L. Davis	"	22
Homer Lewis	June 30, 1868	24
Justus A. Traut	Oct. 6, 1868	28
L.L. Davis	Nov. 21, 1871	33
Justus A. Traut	July 2, 1872	34
Edwin A. Stratton +	July 16, 1872	35
Martin Wilcox	June 6, 1876	40
L.L. Davis	Dec. 5, 1882	51
L.L. Davis	Nov. 20, 1883	59
L.L. Davis	April 29, 1884	62
Albert D. Goodell	Dec. 30, 1884	67
Edwin A. Stratton	May 22, 1888	91
Albert D. Goodell	Oct. 16, 1888	95
Justus A. Traut	Feb. 18, 1890	100
Justus A. Traut	March 25, 1890	102
William M. Morton	Dec. 22, 1896	126
Justus A. Traut	May 8, 1906	163
James M. Burdick	Jan. 14, 1913	185
John Scott +	May 13, 1913	186
Edmund A. Schade	Jan. 9, 1917	201
Harris J. Cook	Nov. 11, 1924	218

Combination Tools

J.A. Littlefield	April 5, 1870	31
George A. Shelly	Nov. 7, 1871	32
Laroy S. Starrett	May 6, 1879	47
Ira Humphrey	May 27, 1879	48
Laroy S. Starrett	Aug. 7, 1883	56
Stephen H. Bellows	March 11, 1884	60
F.W. Ritchie	Aug. 19, 1884	65

Special Purpose Levels

Miscellaneous Purpose or Invention